ENGLISH-SPANISH

JOBSITE

Phrasebook

Kent Shepard

T0206739

BuilderBooks.com
BOOKS THAT BUILD YOUR BUSINESS

A Service of
NAHB
NATIONAL ASSOCIATION
OF HOME BUILDERS

Jobsite Phrasebook

BuilderBooks, a Service of the National Association of Home Builders

Courtenay S. Brown	Director of Book Publishing
Natalie C. Holmes	Editor
Torrie Singletary	Production Editor
E Design Communications	Cover Design
McNaughton-Gunn, Inc.	Printing

Gerald M. Howard	NAHB Executive Vice President and CEO
Mark Pursell	NAHB Senior Staff Vice President, Marketing & Sales Group
Lakisha Campbell	NAHB Staff Vice President, Publication & Affinity Programs

Disclaimer

This publication provides accurate information on the subject matter covered. The publisher is selling it with the understanding that the publisher is not providing legal, accounting, or other professional service. If you need legal advice or other expert assistance, obtain the services of a qualified professional experienced in the subject matter involved. Reference herein to any specific commercial products, process, or service by trade name, trademark, manufacturer, or otherwise does not necessarily constitute or imply its endorsement, recommendation, or favored status by the National Association of Home Builders. The views and opinions of the author expressed in this publication do not necessarily state or reflect those of the National Association of Home Builders, and they shall not be used to advertise or endorse a product.

Printed in the United States of America

11 10   4 5

ISBN-10 0-86718-538-4
ISBN-13 978-0-86718-538-6

   Library of Congress Cataloging-in-Publication Data

Jobsite phrasebook.-- English-Spanish.
   p. cm.
 ISBN-10 0-86718-538-4 (pbk.)
 ISBN-13 978-0-86718-538-6
 1. Building--Terminology. 2. English language--Terms and
phrases--Spanish. 3. Spanish language--Terms and phrases--English. I.
National Association of Home Builders (U.S.)

TH9.J575 2002
690'.01'4--dc21
           2002154928

For further information, please contact:

National Association of Home Builders
1201 15th Street, NW
Washington, DC 20005-2800
800-223-2665
Visit us online at www.BuilderBooks.com.

ABOUT THE AUTHOR

After moving from Colorado to Southern California in the early 1970s, Kent Shepard spent 30 years working as a tradesman and contractor on residential and light commercial projects. These included housing tracts of up to 200 homes, custom homes up to 17,000 square feet, apartment complexes to 1200 units, hotels (look at the octagon and dome on the roof of the "Hotel California" CD by the Eagles), and a wide variety of remodeling projects.

Living again in Colorado near the continental divide, Kent has worked as a framing and finish carpenter, Home Inspector and framing and remodel contractor. He pursues a variety of interests including operating a home inspection and snow removal businesses, photography, etching sandstone, mushroom hunting, skeletal rearticulation (an ostrich project currently in progress), and fatherhood.

Kent is aided in the latter-most endeavor by his son Hunter and daughter Christina. Other books by Kent Shepard are Concrete Jobsite Phrasebook and Framing Jobsite Phrasebook, also published by BuilderBooks.com.

ACKNOWLEDGMENTS

I would like to thank, first of all, Theresa Minch and Jessica Poppe from BuilderBooks for their efforts, which were key in bringing a ten-year project to fruition.

Also Christine Charlip, Torrie Singletary, and Karel Leon for help with the 1st reprint.

My son Hunter helped with the manuscript, and along with my daughter Christina, put up with me while the work was in progress.

Thanks also to Ryan Kramer, my twelve-year-old neighbor who, when my computer became infected with a virus and crashed a few weeks before the final deadline, rescued the files.

BOOK PRODUCTION

BuilderBooks expresses its appreciation to Lesley Foster, Jenny Lambert, and Marvyn Bacigalupo, American Translators Association Accredited Translator and managing editor of *El Nuevo Constructor*.

CONTENTS

Chapter 6: ROOF FRAMING

Chapter 7: ROOFING

ILLUSTRATIONS

PRONUNCIATION

Jobsite Phrasebook uses a simplified method of phrase pronunciation in which the syllable stressed is capitalized. Although a close study of the rules of Spanish grammar is beyond the scope of this book, the following short study should be of some service.

VOWELS:

a- Always an "ah" sound as in "mama" and "mambo"
e- Always a long "a" sound as in "they" and "prey"
i- Always a long "e" sound as in "police" and "piece"
o- Always a long "o" sound as in "oak" and "old"
u- Always a long "u" sound as in "chute" or "tube"

CONSONANTS:

Pronounced as in English: b, ch, d, f, h, k, l, m, n, p, s, and t

c- Sounds like "k" as in "can" and "could"
g- Sounds hard like "gain" and "go" except before "i" and "e" when it is soft like "h" in "hit" or "hey"
j- Sounds like "h" as in "hard" and "hello"
ll- Sounds like "y" as in "yes" and "yard"
n- Pronounced as in English
ñ- Sounds like "ny" and in "canyon". This "n" should have an accent mark above it
qu- Sounds like "k" as in "quiet" and "quality"
r- Pronounced as in English except after n, l, or s at the beginning of a word, when it is rolled (shown by three "r's" in the text pronunciation as in "rrroll")
rr- Is always rolled
v- Sound like "b" as in "banana" or "bill"
w- Seldom used
x- Pronounced as in English
y- Prounced as in the English "yes" and "yard", except where it stands alone, when it is pronounced as a long e
z- Prounced as the English "s"

NOUNS:

Nouns are slightly more complicated and you don't have to know this section to use this book effectively.

In Spanish, nouns are divided into masculine and feminine and generally follow these rules.

MASCULINE NOUNS:

1. The names of male beings (men, uncles, kings).
2. Words ending in "o"
3. Days of the week, months, rivers, oceans and mountains.
4. Nounds ending in "l", "r", or "ma"

FEMININE NOUNS:

1. Names of female beings (women, aunts, queens).
2. Nouns ending in "a"
3. Nouns ending in "ión", "d", "z" or "umbre"

VERBS:

Frequently throughout *Jobsite Phrasebook* the infinitive imperative form of the verb is used for the sake of directness and clarity, and also to avoid problems of propriety in terms of both the singular familiar and polite verb forms denoting "you", which in Spanish can be done.

INTRODUCTION

Jobsite Phrasebook is designed to help improve workplace safety and efficiency by improving communication between builders and their non-English speaking Hispanic employees.

In order to provide very detailed phrases, but still keep the book short enough to allow phrases to be found quickly, the book has been limited to five trades.

The phrases are based on the author's thirty years experience in the building trades and are designed to follow the Concrete, Framing, Roofing, Insulation, and Drywall trades chronologically through the construction process, starting with the foundation and ending with the interior ready for paint.

The operations covered in each trade, in addition to being in chronological order, follow production methods developed by contractors and tradesmen over the years since World War II to speed production, cut costs, and simplify bidding. As much as possible, the material has been written and arranged for use by contractors ranging from those employing large production residential and light commercial crews to small-crew custom home builders. The detail with which each operation is covered also allows it to serve, to a degree, as an instructional text.

In order to speed the process of finding phrases, a format is used that allows the reader to quickly scan the subject matter of an entire page, or find phrases on specific subjects easily through the extensive index.

Safety phrases specific to each trade are included as a group for quick reference.

HIRING

Tools

Do you have tools?
¿Tiene las herramientas?
TYEH-neh lahs eh-rrrah-MYEHN-tahs

Transportation

Do you have transportation?
¿Tiene transportación?
TYEH-neh trahns-pohr-tah-SYOHN

Wages

I will pay you (Ten) dollars per hour.
Le pagaré (diez) dólares por hora.
leh pah-GAH-reh (dyehs) doh-LAHRS pohr OH-rah

We pay by the job.
Pagamos por cada trabajo.
pah-GAH-mohs pohr KAH-dah trah-BAH-hoh

This pays (one hundred ten) dollars.
Este paga (ciento diez) dólares.
EH-steh pah-GAH (SYEHN-toh dyehs) doh-LAHRS

Work hours

We work (eight to four-thirty).
Trabajamos de (ocho a cuatro y media).
trah-bah-HAH-mohs deh (OH-choh ah
KWAH-troh ee MEH-dyah)

You must work late (today / tomorrow).
Debe trabajar hasta tarde (hoy / mañana).
DEH-beh trah-bah-HAHR AH-stah
TAHR-deh (OH-ee / mahn-YAH-nah)

You must finish before you leave.
Debe terminar antes de irse.
DEH-beh tehr-mee-NAHR AHN-tehs deh EER-seh

Payday

I pay every (Friday / other Friday).
Yo pago cada (viernes / otro viernes).
yoh PAH-goh KAH-dah (BYEHR-nehs / OH-troh BYEHR-nehs)

...on the first and fifteenth of the month
...el primero y el quince de cada mes
...ehl pree-MEH-roh ee ehl KEEN-seh deh KAH-dahmehs

Starting date

Can you start now?
¿Puede empezar ahora?
PWEH-deh ehm-peh-SAHR ah-OH-rah

You start (tomorrow / Monday).
Usted empieza (mañana / el lunes).
oo-STEHD ehm-PYEH-sah (mahn-YAH-nah / ehl LOO-nehs)

Come back (Wednesday / in a week).
Regrese (el miércoles / en una semana).
rrreh-GREH-seh (ehl MYEHR-koh-lehs /
ehn OO-nah seh-MAH-nah)

Not hiring

I don't have work for you.
No tengo trabajo para usted.
no TEHN-goh trah-BAH-hoh PAH-rah oo-STEHD

PAPERWORK AND IMMIGRATION

Name

What is your name?
¿Cómo se llama?
KOH-moh seh YAH-mah

My name is (Big Bob).
Me llamo (Big Bob).
meh YAH-moh (Big Bob)

Number

What is your telephone number?
¿Cuál es su número de teléfono?
kwahl ehs soo NOO-meh-roh deh teh-LEH-foh-noh

Filling out forms

Fill out (this / these) form(s).
Llene esta formularios.
YEH-neh (EH-stay / EH-stohs) fohr-moo-LAHR-yoh(s)

Have some one help you.
Conseguir ayuda.
kohn-seh-GEER ah-YOO-dah

Federal tax form	**This is a federal tax form.** Este es un formulario de impuestos federales. *EH-steh ehs oon fohr-moo-LAHR-yoh deh eem-* *PWEH-stohs feh-deh-RAH-lehs*
State tax form	**This is a state tax form.** Este es un formulario de impuestos estatales. *EH-steh ehs oon fohr-moo-LAHR-yoh deh eem-PWEH-stohs* *eh-stah-TAH-lehs*
Immigration form	**This is an immigration form.** Este es un formulario de inmigración. *EH-steh ehs oon fohr-moo-LAHR-yoh* *deh ee-mee-grah-SYOHN*
Identification	**I need to see your identification.** Necesito ver su identificación. *neh-seh-SEE-toh behr soo ee-dehn-tee-fee-kah-SYOHN*

It must be the one you list on the immigration form.
Debe ser la misma que puso en el formulario de inmigración.
DEH-beh sehr lah MEES-mah keh POO-soh ehn ehl fohr-
moo-LAHR-yoh deh ee-mee-grah-SYOHN

Do you have a driver's license?
¿Tiene una licencia de conducción?
TYEH-neh OO-nah lee-SEHN-syah deh kohn-dook-SYOHN

Without identification, I can't hire you.
Sin identificación, no lo puedo contratar.
seen ee-dehn-tee-fee-kah-SYOHN, noh loh
PWEH-doh kohn-trah-TAHR

FIRING

You are fired.
Está despedido.
EH-stah dehs-peh-DEE-doh

Turn in your tools.
Devuelva sus herramientas.
de-BWEHL-bah soos eh-rrrah-MYEHN-tahs

Leave the job site (now / within thirty) minutes.
Salga del área de trabajo (ya / antes de treinta) minutos.
SAHL-gah de ah-REH-ah deh trah-BAH-hoh
(yah / AHN-tehs deh TRAIN-tah mee-NOO-tohs)

Lay-off

I have to lay you off.
Tengo que darle lay off.
TEHN-goh keh DAHR-leh lay off

There is no more work for you.
No hay más trabajo para usted.
no AH-ee mahs trah-BAH-hoh PAH-rah oo-STEHD

Last paycheck

Go to the office for your check.
Vaya a la oficina por su cheque.
BAH-yah ah lah oh-fee-SEE-nah pohr soo CHEH-keh

Come with me.
Venga conmigo.
BEHN-gah kohn-MEE-goh

PAYROLL

Time sheets

Give me your time sheet.
Deme su hoja de tiempo.
DEH-meh soo HOH-hah deh TYEHM-poh

Your addition is wrong.
Su suma está mal.
soo SOO-mah EH-stah mahl

Missed work

You were not here that day.
Usted no estuvo aquí ese día.
oo-STEHD noh eh-STOO-boh ah-KEE EH-seh DEE-ah

You left early that day.
Usted se fue temprano ese día.
oo-STEHD seh fweh tehm-PRAH-noh EH-seh DEE-ah

You were on another job that day.
Usted estaba en otro trabajo ese día.
oo-STEHD eh-STAH-bah ehn OH-trah
trah-BAH-hoh EH-seh DEE-ah

GENERAL SAFETY

Safety Meetings

**We have a safety meeting
(every Monday morning / now).**
Tenemos una reunión de seguridad
(cada lunes por la mañana / ahora).
teh-*NEH-mohs OO-nah rrreh-oon-YOHN deh
seh-goo-ree-DAHD (KAH-dah LOO-nehs pohr
lah mahn-YAH-nah / ah-OH-rah)*

Go to (my truck / the trailer).
Vaya (a mí camión / al tráiler).
Bay-yah-(ah mee kahm-YOHN / ahl trailer)

Safety information

This is safety information.
Ésta es información de seguridad.
EH-stah is een fohr mah-SYOHN deh seh-goo-ree-DAHD

Read this.
Lea esto.
LEH-ah ehs-STOH

Reporting Accidents

**You must report accidents
(immediately / within four days).**
Debe reportar los accidentes
(inmediatamente / antes de cuatro días).
*DEH-beh rrreh-pohr-TAHR lohs ahk-see-DEHN-tehs
(ee-mee-dyah-tah-MEHN-teh / AHN-tehs deh
KWAH-troh DEE-ahs)*

Report accidents to (me / your supervisor).
Reporte los accidentes a (mí / su supervisor).
*rrreh-POHR-teh lohs ahk-see-DEHN-tehs ah
(mee / soo soo-pehr-bee-SOHR)*

Protective gear

You must wear (safety glasses / a hard hat).
Debe usar (gafas de seguridad / casco duro).
*DEH-beh oo-SAHR (GAH-fahs deh
seh-goo-ree-DAHD / KAHS-koh DOO-roh)*

…gloves / a dust mask / a respirator / ear plugs
…guantes / máscara anti-polvo / respirador /
tapones de oído
*…GWAHN-tehs / mah-SKAH-rah AHN-tee POHL-boh
/ rehs-peer-ah-DOHR / tah-POH-nehs de oh-EE-doh*

Scaffolding

Stay off the scaffolding.
No se monte en el andamiaje.
noh seh MOHN-teh ehn ehl ahn-dah-MYAH-heh

Install guardrails here.
Instalar barandas aquí.
een-stah-LAHR bah-RAHN-dahs ah-KEE

(Forty-two) inches to the top.
A (cuarenta y dos) pulgadas del tope.
ah (kwah-REHN-tah ee dohs) pool-GAH-dahs dehl TOH-peh

Fall protection

You must wear a harness.
Debe usar arreos.
DEH-beh oo-SAHR ah-RRREH-ohs

You must tie off.
Debe amarrarse.
DEH-beh ahr-mah-RRRAHR-seh

Use the safety rope.
Usar la cuerda de seguridad.
oo-SAHR lah KWEHR-dah de seh-goo-ree-DAHD

Toilet use

Keep the toilet clean.
Mantenga limpio el baño.
mahn-TEHN-gah LEEM-pyoh ehl BAHN-yoh

Ladders

Set up the ladder.
Colocar la escalera.
koh-loh-KARH lah eh-skah-LEH-rah

Put it (here / there).
Ponerla (aquí / allá).
poh-NEHR-lah (ah-KEE / ah-YAH)

Make it safe.
Asegurarla.
ah-seh-goo-RAHR-lah

Tying off

Use the safety rope.
Usar la cuerda de seguridad.
oo-SAHR lah KWEHR-dah deh seh-goo-ree-DAHD

Nail this to the peak.
Clavarla a la cima.
klah-BAHR-lah ah lah SEE-mah

Working conditions

Stop working.
Parar de trabajar.
pah-RAHR deh trah-bah-HAHR

It is too (windy / rainy / cold) to work.
Es demasiado (ventoso / lluvioso / frío) para trabajar.
*ehs deh-mah-SYAH-doh (behn-TOH-soh /
yoo-BYOH-soh / FREE-oh) PAH-rah trah-bah-HAHR*

CONCRETE

Rebar impalement

Put the caps on the rebar.
Poner tapas en la barra de refuerzo.
*poh-NEHR TAH-pahs ehn lah BAH-rrrah
deh rrreh-FWEHR-soh*

Bend over the rebar.
Doblar la barra de refuerzo.
doh-BLAHR lah BAH-rrrah deh rrreh-FWEHR-soh

Trenches

Stay out of the trenches.
Mantenerse fuera de las zanjas.
mahn-teh-NEHR-seh FWEHR-rah deh lahs SAHN-hahs

Shore the trenches.
Apuntalar las zanjas.
ah-poon-tah-LAHR lahs SAHN-hahs

Cover the trench (here / all the way around).
Cubrir la zanja (aquí / todo alrededor).
koo-BREER lah SAHN-hah (ah-KEE / TOH-doh
ahl-rrreh-deha-DOHR).

Scaffolding

Stay off the scaffolding.
No montarse en el andamiaje.
noh mohn-TAHR-seh ehn ehl ahn-dah-MYAH-heh

Brace the scaffolding.
Arriostrar el andamiaje.
ah-*rrree-oh-STRAHR ehl ahn-dah-MYAH-heh*

Guardrails

Put guardrails here.
Poner barandas aquí.
poh-NEHR bah-rahn-DAHS ah-KEE

Make them (forty-two) inches to the top.
Que queden a (cuarenta y dos) pulgadas del tope.
keh keh-DEHN ah (kwah-REHN-tah ee dohs)
pool-GAH-dahs dehl TOH-peh

Put a rail in the middle.
Poner una baranda a la mitad.
poh-NEHR OO-nah bah-RAHN-dah ah lah mee-TAHD

FRAMING

Saw safety

Don't put your fingers behind the blade.
No poner los dedos detrás de la cuchilla.
no poh-NEHR lohs DEH-dohs deh-TRAHS
deh lah koo-CHEE-yah

Don't wedge the guard.
No cuñar la protección.
no koon-YAHR lah proh-tek-SYON

Wall bracing

Put more braces on (that wall / the walls).
Poner más riostras en (esa pared / las paredes).
poh-NEHR mahs rrree-OH-strahs ehn
(EH-sah pah-REHD / lahs pah-REH-dehs)

High winds are predicted.
Se predicen vientos fuertes.
seh preh-dee-SEHN BYEHN-tohs FWEHR-tehs

Electrical issues

That cord is unsafe, (don't use it / replace it).
Ese cable es peligroso, (no usarlo /reemplazarlo).
EH-seh KAH-bleh ehs peh-lee-GROH-soh,
(noh oo-SAHR-loh / rrreh-ehm-plah-SAHR-loh)

Keep the plugs out of the water.
Mantener los enchufes fuera del agua.
mahn-teh-NEHR lohs ehn-CHOO-fehs
FWEH-rah dehl AH-gwah

Unplug that cord.
Desconectar ese cable.
dehs-koh-nehk-TAHR EH-seh KAH-bleh

Ladder safety

That ladder is unsafe, don't use it.
Esa escalera es peligroso, no usarla.
EH-sah eh-skah-LEH-rah ehs peh-lee-GROH-soh,
noh oo-SAHR-lah

Set the ladder up more carefully.
Posicionar la escalera con más cuidado.
poh-see-syoh-NAHR lah eh-skah-LEH-rah
kohn mahs kwee-DAH-doh

Fall prevention

Cover that hole.
Cubrir ese hueco.
koo-BREER EH-seh WEH-koh

Use two by (four / six) and plywood.
Usar dos por (cuatro / seis) y madera terciada.
oo-SAHR dohs pohr (KWAH-troh / sais)
ee mah-DEH-rah tehr-see-AH-dah

Eye and ear protection

Put on (safety glasses / ear protection).
Ponerse (gafas de seguridad / protección para los oídos).
poh-NEHR-seh (GAH-fahs deh seh-goo-ree-DAHD /
proh-tehk-SYOHN PAH-rah lohs oh-EE-dohs)

ROOFING

Fall prevention

You must (wear a harness / tie off to the roof).
Debe (usar arreos / amarrarse al techo).
DEH-beh (oo-SAHR ah-REH-ohs /
ahr-mah-RRRAHR-seh ahl TEH-choh)

Stay off the roof, it is too slippery.
No montarse en el techo, es demasiado resbaladizo.
noh mohn-TAHR-seh ehn ehl TEH-choh,
ehs deh-mah-SYAH-doh rrrehs-bah-lah-DEE-soh

**Injury from
thrown scrap**

Before you throw scrap, (look / yell "headache!").
Antes de botar los desperdicios, (mire / grite "¡headache!").
AHN-tehs deh boh-TAHR lohs dehs-pehr-DEE-syoh
(MEE-reh / GREE-teh "headache!")

INSULATION AND DRYWALL

Clothing

Wear long sleeves.
Usar mangas largas.
oo-SAHR MAHN-gahs LAHR-gahs

Eye protection

Do you have safety glasses?
¿Tiene gafas de seguridad?
TYEH-neh GAH-fahs deh seh-goo-ree-DAHD

Wear safety glasses.
Usar gafas de seguridad.
oo-SAHR GAH-fahs deh seh-goo-ree-DAHD

Respirator

Do you have dust masks?
¿Tiene máscaras anti-polvo?
TYEH-neh mah-SKAH-rahs AHN-tee POHL-boh

Wear a (respirator / dust mask).
Usar (respirador / máscara anti-polvo).
oo-SAHR (rehs-pee-rah-DOHR /
mah-SKAH-rah AHN-tee POHL-boh)

Get the masks from my truck.
Traer las máscaras de mi cami.
oo-SAHR (rehs-pee-rah-DOHR /
mah-SKAH-rah AHN-tee POHL-boh

COMPRESSOR

et-up

Set up the compressor.
Armar el compresor.
ahr-MAHR ehl kohm-preh-SOHR

It must be level.
Debe quedar nivelado.
DEH-beh keh-DAHR nee-beh-LAH-doh

lectrical

Before you plug it in, turn off the switch.
Antes de conectarlo, apagar el interruptor.
AHN-tehs deh koh-nehk-TAHR-lo,
ah-pah-GAHR ehl een-teh-RRRuhp-tohr

Replace the plug.
Reemplazar el enchufe.
rrreh-ehm-plah-SAHR ehl ehn-CHOO-feh

The switch is bad.
El interruptor está mal.
ehl een-teh-RRRuhp-tohr EH-stah MAHL

Unplug it.
Desconectarlo.
dehs-koh-nehk-TAHR-loh

peration

Close the drain.
Cerrar el desagüe.
seh-RRRAHR ehl ehs-AH-gweh

(Turn on / plug in) the compressor.
(Prender / conectar) el compresor.
(prehn-DEHR / koh-nehk-TAHR) ehl kohm-preh-SOHR

Turn off the compressor.
Apagar el compresor.
ah-pah-GAHR ehl kohm-preh-SOHR

Turn the pressure (up / down).
(Subir / bajar) la presión.
(soo-BEER / bah-HAHR) lah preh-SYOHN

Problems

Is it (plugged in / turned on)?
¿Está (conectado / prendido)?
eh-STAH (koh-nehk-TAH-doh / prehn-DEE-doh)

Check the breaker.
Revisar el cortacircuitos.
reh-bee-SAHR ehl Kohr-tah-seer-KWEE-tohs

Check the reset button.
Revisar el botón de reposición.
reh-bee-SAHR ehl boh-TOHN de reh-poh-see-SYOHN

Shake the cord.
Sacudir el cable.
sah-koo-DEER ahl KAH-bleh

Draining the tank

Drain the tank (every night).
Vaciar el tanque (cada noche).
bah-SYAHR ehl TAHN-keh (KAH-dah NOH-cheh)

Checking the oil

Check the oil.
Revisar el aceite.
rrreh-bee-SAHR ehl ah-seh-EE-teh

Change the oil.
Cambiar el aceite.
kahm-BYAHR ehl ah-seh-EE-teh

Use the compressor oil.
Usar aceite de compresor.
oo-SAHR ah-seh-EE-teh deh kom-preh-SOHR

Don't over-fill it.
No sobrellenar.
noh soh-breh-yeh-NAHR

NAIL GUN

iling

Oil the gun (every day / now).
Aceitar la clavadora (todos los días / ahora).
ah-seh-ee-TAHR lah klah-bah-DOH-rah
(TOH-dohs lohs dyehs / ah-OH-rah)

Use (three) drops.
Usar (tres) gotas.
oo-SAHR (trehs) GOH-tahs

Clearing jams

Disconnect the hose.
Desconectar la manguera.
dehs-koh-nehk-TAHR lah mahn-GEH-rah

Don't point it at your face.
No la apunte hacia su cara.
noh lah ah-POON-teh AH-syah soo KAH-rah

Remove (the front / the nail).
Remover (el frente / el clavo).
rrreh-moh-BEHR (ehl FREHN-teh / ehl KLAH-boh)

Replace the front.
Reemplazar el frente.
rrreh-ehm-plah-SAHR ehl FREHN-teh

HOSES

eaking

Tighten the clamp.
Apretar la abrazadera.
ah-preh-TAHR lah ah-brah-sah-DEH-rah

Replace this fitting.
Reemplazar este accesorio.
rrreh-ehm-plah-SAHR EH-steh ahk-seh-SOH-ree-yoh

Repairing holes

Find the hole.
Encontrar el agujero.
ehn-kohn-TRAHR ehl Ah-goo-HEH-roh

Cut the hose.
Cortar la manguera.
kohr-TAHR lah mahn-GEH-rah

Install the fitting.
Instalar el accesorio.
een-stah-LAHR ehl ahk-seh-SOH-ree-yoh

Tighten the clamp.
Apretar la abrazadera.
ah-preh-TAHR lah ah-brah-sah-DEH-rah

Rolling out / up

(Unroll / plug in) the hoses.
(Desenrollar / conectar) las mangueras.
(dehs-ehn-rroh-YAHR / koh-nehk-TAHR)
lahs mahn-GEH-rahs

(Coil / braid) the hose.
(Enrollar / trenzar) la manguera.
(ehn-rroh-YAHR / trehn-SAHR)_lah mahn-GEH-rah

POWER TOOLS

Cleaning

Clean it.
Limpiarla.
leem-PYAHR-lah

Use the blow gun.
Usar la sopladora.
oo-SAHR lah soh-plah-DOH-rah

Electrical

Replace the (cord / plug).
Reemplazar (el cable / enchufe).
rrreh-ehm-plah-SAHR (ehl KAH-bleh / ehn-CHOO-feh)

Fix the (cord / plug).
Arreglar (el cable / enchufe).
ah-rrreh-GLAHR (ehl KAH-bleh / ehn-CHOO-feh)

Use a (bigger / shorter) cord.
Usar un cable más (grande / corto).
oo-SAHR oon KAH-bleh mahs (GRAHN-deh / KOHR-toh)

Changing oil

Change the oil.
Cambiar el aceite.
kahm-BYAHR ehl ah-seh-EE-teh

It takes (ninety) weight oil.
Se necesita aceite de peso (noventa).
seh neh-seh-SEE-tah ah-SH-teh deh
PEH-soh (noh-BEHN-tah)

Put it in here.
Ponerlo aquí.
poh-NEHR-loh ah-KEE

Blade replacement

Replace the blade.
Reemplazar la cuchilla.
rrreh-ehm-plah-SAHR lah koo-CHEE-yah

Use a (wood, metal) blade.
Usar una cuchilla de (madera, metal).
oo-SAHR OO-nah koo-CHEE-yah deh
(mah-DEH-rah / meh-TAHL)

...masonry / abrasive / steel / carbide
...albañilería/ abrasiva / acero / carburo
...ahl-bahn-yee-LEHR-yah / ah-brah-SEE-bah/
ah-SEH-roh / kahr-BOO-roh

BATTER BOARDS

**Building
batter boards**

Build the batter boards.
Construir las tablas para las cuerdas.
kohn-stroo-EER lahs TAH-blahs PAH-rah lahs KWEHR-dahs

Put it here.
Ponerla aquí.
poh-NEHR-lah ah-KEE

FORM SETTING (FOOTINGS AND SLAB)

Unloading

Unload the forms.
Descargar los moldes.
dehs-kahr-GAHR lohs MOHL-dehs

Use (four) men.
Usar (cuatro) hombres.
oo-SAHR (KWAH-troh) OHM-brehs

Get some help.
Conseguir ayuda.
kohn-seh-GEER ah-YOO-dah

Form material

Use two by (four / six / eight / ten) for forms.
Usar el dos por (cuatro / seis / ocho / diez) para los moldes.
*oo-SAHR ehl dohs pohr (KWAH-troh / sais / OH-choh
/ dyehs) PAH-rah lohs MOHL-dehs*

Laying out stakes

Lay out the stakes.
Colocar las estacas.
koh-loh-KAHR lahs eh-STAH-kahs

Use the (metal / wood) stakes.
Usar estacas de (metal / madera).
oo-SAHR eh-STAH-kahs deh (meh-TAHL / mah-DEH-rah)

Figure 1. Batter Boards
          Tablas para las cuerdas

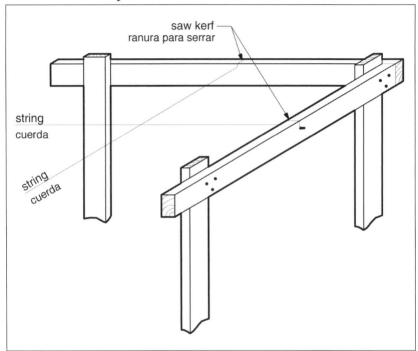

Typical Batter Board construction.

Laying out braces	**Lay out the braces.** Colocar las riostras. *koh-loh-KARH lahs rrree-OH-strahs*
Brace size	**Cut braces out of the two by four, (fourteen) foot.** Cortar las riostras de los dos por cuatro, (catorce) pies. *kohr-TAHR lahs rrree-OH-strahs deh lohs dohs pohr* *KWAH-troh, (kah-TOHR-seh) pyehs*
Setting forms	**Set the forms.** Erigir los moldes. *eh-reeh-HEER lohs MOHL-dehs*
	Do you know how to set forms? ¿Sabe erigir los moldes? *SAH-beh eh-ree-HEER lohs MOHL-dehs*
Form nails	**Use duplex nails.** Usar clavos duplex. *oo-SAHR KLAH-bos DOO-plehks*
	The nails are in (my truck / the box). Los clavos están en (mi camión / la caja). *lohs KLAH-bohs EH-stahn ehn* *(mee kahm-YOHN / lah KAH-hah)*
Form location	**The string is on the inside of the forms.** La cuerda esta el interior de los moldes. *lah KWEHR-dah ehl eh-STAH ehn een-tehr-YOHR* *deh lohs MOHL-dehs*
	The inside of the form goes here. El interior de los moldes va aquí. *ehl een-tehr-YOHR deh lohs MOHL-dehs bah ah-KEE*
Footing dimensions	**Make it (sixteen / twenty-four) inches wide.** Debe tener (dieciséis / veinticuatro) pulgadas de ancho. *DEH-beh teh-NEHR (dyehs-ee-SAIS / byehn-tee-KWAH-troh)* *pool-GAH-dahs deh AN-cho*

Make it (thirty-five) (feet / inches) long.
Que tenga (treinta y cinco) (pies / pulgadas) de largo.
keh TEHN-gah (TRAIN-tah ee SEEN-koh)
(pyehs / pool-GAH-dahs) deh LAHR-goh

It steps (up / down) (twenty-four) inches here.
(Sube / baja) (veinte y cuatro) pulgadas aquí.
(SOO-beh / BAH-hah) (BAIN-teh ee KWAH-troh)
pool-GAH-dahs ah-KEE

Squaring corners (3-4-5 method)

The corner must be square.
La esquina debe ser cuadrada.
lah eh-SKEE-nah DEH-beh sehr kwah-DRAH-dah

Mark this side at (three / four) feet.
Marcar este lado a (tres / cuatro) pies.
mahr-KAHR EH-steh LAH-doh ah (trehs / KWAH-troh) pyehs

This should measure (five) feet.
Éste debe medir (cinco) pies.
EH-steh DEH-beh meh-DEER (SEEN-koh) pyehs

Leveling forms

Level the forms.
Nivelar los moldes.
nee-beh-LAHR lohs MOHL-dehs

It must be level.
Debe estar nivelado.
DEH-beh eh-STAHR nee-beh-LAH-doh

Bracing forms

Brace the forms.
Arriostrar los moldes.
ah-rree-oh-STRAHR lohs MOHL-dehs

It must be straight.
Debe estar derecho.
DEH-beh eh-STAHR deh-REH-choh

It's not (straight / level).
No está (derecho / nivelado).
noh EH-stah (deh-REH-choh / nee-beh-LAH-doh)

Stripping forms	**(Strip / clean / oil) the forms.**
	(Pelar / limpiar / aceitar) los moldes.
	(peh-LAHR / leem-PYAHR / ah-seh-ee-TAHR) lohs MOHL-dehs
	Load them on the truck.
	Cargarlos en el camión.
	kahr-GAHR-lohs ehn ehl kahm-YOHN

SLAB-ON-GRADE (SURFACE PREPARATION)

Laying out sand	**Lay out the sand.**
	Extender la arena.
	ehs-sten-DEHR lah ah-REH-nah
	Make it (three / four) inches deep.
	Que tenga (tres / cuatro) pulgadas de profundidad.
	keh-TEHN-gah (trehs / KWAH-troh)
	pool-GAH-dahs deh proh-foon-dee-DAHD
	Wet the sand.
	Mojar la arena.
	moh-HAHR lah ah-REH-nah

Vapor barrier	**Lay out the plastic.**
	Extenderel plástico.
	ehs-sten-DEHR ehl PLAH-stee-koh
	Overlap the edges twelve inches.
	Traslapar los bordes doce pulgadas.
	trahs-lah-PAHR lohs BOHR-dehs DOH-seh pool-GAH-dahs

Welded wire	**Lay out the welded wire.**
	Extender el cable soldado.
	ehs-sten-DEHR ehl KAH-bleh sohl-DAH-doh
	Overlap it (two) meshes.
	Traslaparlo (dos) vueltas.
	trahs-lah-PAHR-loh (dohs) BWEHL-tahs
	It must be two inches off the ground.
	debe quedar dos pulgadas del suelo.
	DEH-beh keh-DAHR dohs pool-GAH-dahs dehl soo-EH-loh

Figure 2. Forming a Slab On Grade
Cimbrado de losa a nivel

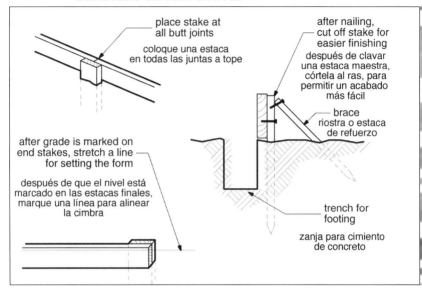

place stake at all butt joints

coloque una estaca en todas las juntas a tope

after nailing, cut off stake for easier finishing

después de clavar una estaca maestra, córtela al ras, para permitir un acabado más fácil

brace
riostra o estaca de refuerzo

after grade is marked on end stakes, stretch a line for setting the form

después de que el nivel está marcado en las estacas finales, marque una línea para alinear la cimbra

trench for footing

zanja para cimiento de concreto

Block-outs

Put a block-out here.
Poner un bloqueo aquí.
poh-NEHR oon bloh-KEH-oh ah-KEE

(Sixty) inches by (thirty-six) inches.
(Sesenta) por (treinta y seis) pulgadas.
(seh-SEHN-tah) pohr (TRAIN-tah ee sais) pool-GAH-dahs

(eight) (feet / inches) (to it / to center).
A (ocho) (pies / pulgadas) (de él / del centro).
Ah (OH-choh) (pyehs / pool-GAH-dahs) (de / dehl SEHN-troh)

(Butt / hook) the form board.
(Tocar / enganchar) la tabla de molde.
(toh-KAHR / ehn-gahn-CHAR) lah TAH-blah deh MOHL-deh

FORM SETTING (WALLS)

Unloading forms

Unload the forms.
Descargar los moldes.
dehs-kahr-GAHR lohs MOHL-dehs

Use (three) men.
Usar (tres) hombres.
oo-SAHR (trehs) OHM-brehs

Get some help.
Conseguir ayuda.
kohn-seh-GEER ah-YOO-dah

Form size

Use (eight) foot forms here.
Usar los moldes de (ocho) pies aquí.
oo-SAHR lohs MOHL-dehs deh (OH-choh) pyehs ah-KEE

Laying out forms

Lay out the forms.
Colocar los moldes.
koh-loh-KAHR lohs MOHL-dehs

Put them (inside / outside) the footings.
Ponerlos (dentro / fuera) de los fundamentos.
poh-NEHR-lohs (DEHN-troh / FWEH-rah)
de lohs foon-dah-MEHN-tohs

Figure 3. Forming Walls
Formación de las paredes

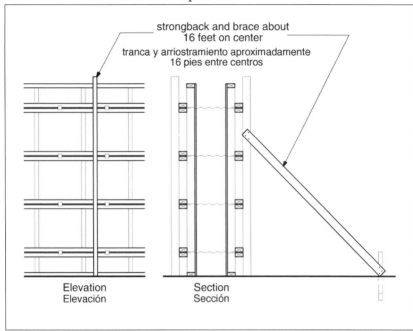

strongback and brace about 16 feet on center

tranca y arriostramiento aproximadamente 16 pies entre centros

Elevation
Elevación

Section
Sección

Oiled side up.
Lado aceitado arriba.
LAH-doh ah-seh-ee-TAH-doh ah-RRREE-bah

aying out braces

Lay out the braces.
Colocar las riostras.
koh-loh-KAHR lahs rrree-OH-strahs

Use (fourteen) footers for braces.
Usar los maderos de (catorce) para las riostras.
oo-SAHR lohs mah-DEH-rohs deh
(kah-TOHR-seh) PAH-rah lahs rrree-OH-strahs

Put a brace every (eight) feet.
Poner una riostra cada (ocho) pies.
poh-NEHR OO-nah rrree-OH-strah
KAH-dah (OH-choh) pyehs

aying out stakes

Lay out the stakes.
Colocar las estacas.
koh-loh-KAHR lahs eh-STAH-kahs

Lay out one per brace.
Esparcir una por cada riostra.
ehs-pahr-SEER OO-nah pohr KAH-dah rrree-OH-strah

etting forms

Set the (inside / outside) forms.
Erigir los moldes (interiores / exteriores).
eh-ree-HEER lohs MOHL-dehs
(een-tehr-YOHR-ehs / ehks-tehr-YOHR-ehs)

Do you know how to set forms?
¿Sabe erigir los moldes?
SAH-beh eh-ree-HEER lohs MOHL-dehs

This is a (four) (foot / inch) wall.
Ésta es una pared de (cuatro) (pies / pulgadas).
EH-stah ehs OO-nah pah-REHD deh (KWAH-troh)
(pyehs / pool-GAH-dahs)

stalling forms

Put the forms on the footing.
Poner los moldes en el fundamento.
poh-NEHR lohs MOHL-dehs ehn ehl foon-dah-MEHN-toh

Nail the edges.
Clavar los bordes.
klah-BAHR lohs BOHR-dehs

Fasten the (horizontal / vertical / diagonal) braces.
Atar las riostras (horizontales / verticales / diagonales).
ah-TAHR lahs rrree-OH-strahs (oh-ree-sohn-TAHL-ehs /
behr-tee-KAHL-ehs / dyah-goh-NAH-lehs)

Make it (one / two / three) high.
Dé una altura de (uno / dos / tres).
deh OO-nah ahl-TOO-ra deh (OO-noh / dohs / trehs)

Form liners

Install the form liners here.
Instalar los alineadores de moldes aquí.
een-stah-LAHR lohs ah-lee-neh-ah-DOH-rehs
deh MOHL-dehs ah-KEE

Bracing forms

Brace the forms.
Arriostrar los moldes.
ah-rree-oh-STRAHR lohs MOHL-dehs

Put braces (in the middle / at the top / bottom).
Poner las riostras (en el medio / arriba / abajo).
poh-NEHR-lahs rrree-OH-strahs
(en ehl MEHD-yoh / ah-RRREE-bah / ah-BAH-hoh)

Nail them to the turnbuckles.
Clavarlas a los tensores.
klah-BAHR-lahs ah lohs tehn-SOHR-ehs

Installing spreaders

Install the spreaders.
Instalar los separadores.
een-stah-LAHR lohs seh-pah-rah-DOHR-rehs

Block-out locations

Block out for a (window / door) here.
Bloquear para una (ventana / puerta) aquí.
bloh-keh-AHR PAH-rah OO-nah
(behn-TAH-nah / PWEHR-tah) ah-KEE

Measuring block-outs	**Make it (eight) inches (wide / high).** Que tenga (ocho) pulgadas de (ancho / alto). *keh TEHN-gah (OH-choh) pool-GAH-dahs deh* *(AHN-choh / AHL-toh)*
	(Fifty) inches to the top. A (cincuenta) pulgadas del tope. *ah (seen-KWEHN-tah) pool-GAH-dahs dehl TOH-peh*
Bracing block-outs	**Brace it like this.** Arriostrar así. *ah-rree-oh-STRAHR ah-SEE*
Setting steel	**Set the steel.** Cuadre el acero. *koh-loh-KAHR ehl ah-SEH-roh*
Installing form ties	**Install the form ties.** Instalar las ataduras de moldes. *een-stah-LAHR las ah-tah-DOO-rahs deh MOHL-dehs*
Plumbing forms	**Plumb the (corners / forms).** Aplomar (las esquinas / los moldes). *ah-ploh-MARH (lahs eh-SKEE-nahs / lohs MOHL-dehs)*
	(Tighten / loosen) the turnbuckle. (Apretar / aflojar) el tensor. *(ah-preh-TAHR / ah-floh-HAHR) ehl tehn-SOHR*
Aligning forms	**Align the forms.** Alinear los moldes. *ah-leen-YAHR lohs MOHL-dehs*
	(Sight / string) the wall. (Visar / cordonear) la pared. *(bee-SAHR / kohr-dohn-YAHR) lah pah-REHD*
	It must be (plumb / straight). Debe estar (aplomada / derecha). *DE-beh eh-STAHR (ah-ploh-MAH-dah / deh-REH-chah)*

Stripping forms	**Strip the forms.** Pelar los moldes. *peh-LAHR lohs MOHL-dehs*
Snapping forms ties	**Snap off the ties.** Soltar las ataduras. *sohl-TAHR lahs ah-tah-DOO-rahs*

HEAVY FORMS (CRANE)

Hooking and unhooking	**This one first.** Éste primero. *EH-steh pree-MEH-roh*
	(Hook / unhook) the form. (Enganchar / desenganchar) el molde. *(ehn-gahn-CHAR / dehs-ehn-gahn-CHAR) ehl MOHL-deh*
	Is the hook attached? ¿Está enlazado el gancho? *EH-stah ehn-lah-SAH-doh ehl GAHN-choh*
	Watch out! ¡Cuidado! *kwee-DAH-doh*
Guiding heavy forms	**Guide the form.** Guiar el molde. *gee-YAHR ehl MOHL-deh*
	Tie on the rope. Atar la cuerda. *ah-TAHR lah KWEHR-dah*
	Turn it this way. Voltearla en este sentido. *bohl-teh-AHR-lah ehn EH-steh sehn-TEE-doh*
Setting heavy forms	**...up / down / in / out** ...arriba / abajo / adentro / afuera *...ah-RRREE-bah / ah-BAH-hoh /* *ah-DEHN-troh / ah-FWEH-rah*

...to me / to you
...hacia mí / hacia usted
...AH-syah mee / AH-syah oo-STEHD

Bracing heavy forms | **Brace the forms.**
Arriostrar los moldes.
ah-rree-oh-STRAHR lohs MOHL-dehs

One brace every (eight) feet.
Una riostra cada (ocho) pies.
OO-nah rrree-OH-strah KAH-dah (OH-choh) pyehs

Put a brace (here / there).
Poner una riostra (aquí / allá).
poh-NEHR OO-nah rrree-OH-strah (ah-KEE / ah-YAH)

STAIRWAYS AND RAMPS

Grading | **Grade this for the (stairway / ramp).**
Graduar para (la escalera / la rampa).
grah-DWAHR PAH-rah (lah eh-skah-LEH-rah /
lah RRRAHM-pah)

Make it (forty) (feet / inches) (wide / long).
Que tenga (cuarenta) (pies / pulgadas) de (ancho / largo).
keh TEHN-gah (kwah-REHN-tah)
(pyehs / pool-GAH-dahs) (deh AHN-choh / LAHR-goh)

Make it (seventy-three) inches high here to nothing here.
Que tenga (setenta y tres) pulgadas de altura aquí a cero aquí.
keh TEHN-gah (seh-TEHN-tah ee trehs) pool-GAH-dahs
deh ahl-TOO-rah ah-KEE ah SEH-ro ah-KEE

Ramp angle | **The ramp angle is (twelve) degrees.**
El ángulo de la rampa es (doce) grados.
ehl ahn-GOO-loh deh lah RRRAHM-pah ehs
(DOH-seh) GRAH-dohs

It rises (three) inches in one foot.
Sube (tres) pulgadas en un pie.
SOO-beh (trehs) pool-GAH-dahs ehn oon p-YEH

Forming stairs	**Form the stairs here.**
	Moldear las escaleras aquí.
	mohl-deh-AHR lohs ehs-KAH-leh-rahs ah-KEE

Stair rise and run	**(The riser / the tread) is (seven / ten) inches.**
	(La contrahuella) es de (siete / diez) pulgadas.
	(Kohn-trah-oo-EH-yah) ehs deh
	(SYEH-teh / dyehs) pool-GAH-dahs

The riser is (plumb / angled) (thirty six) degrees.
La contrahuella está (aplomada / inclinada)
a (treinte y seis) grados.
lah kohn-trah-oo-EE-yah EH-stah (ah-ploh-MAH-dah
/ een-klee-NAH-dah) ah (TRAIN-tah ee sais) GRAH-dohs

■■■■■■■■■ FLATWORK AND DRIVEWAYS GRADING

Dig out for the walk.
Excavar para la acera.
ehks-kah-BAHR PAH-rah lah ah-SEH-rah

Dig between the strings.
Excavar entre las cuerdas.
ehks-kah-BAHR EHN-treh lahs KWEHR-dahs

Make it (eight) inches below the string.
Que tenga (ocho) pulgadas por debajo de la cuerda.
keh TEHN-gah (OH-choh) pool-GAH-dahs pohr
deh-BAH-hoh deh lah KWEHR-dah

It must be (twenty-two) (feet / inches) wide.
Debe tener (veinte y dos) (pies / pulgadas) de ancho.
DEH-beh teh-NEHR (BAIN-teh ee dohs)
(pyehs / pool-GAH-dahs) deh AHN-choh

It must be flat.
Debe estar plano.
DEH-beh eh-STAHR PLAH-noh

Void forms	**Put the void forms in the driveway.**
	Poner las cajas en la entrada.
	poh-NEHR lahs KAH-hahs ehn lah ehn-TRAH-dah

SETTING STEEL

Rebar size

Use number (four / five / six / eight) rebar here.
Usar la barra de refuerzo número
(cuatro / cinco / seis / ocho) aquí.
oo-SAHR lah BAH-rrrah deh rreh-FWEHR-soh NOO-meh-roh
(KWAH-troh / SEEN-koh / sais / OH-choh) ah-KEE

Direction of run

They run (this way / both ways).
Van en (este sentido / ambos sentidos).
bahn ehn (EH-steh sehn-TEE-doh /
AHM-bohs sehn-TEE-dohs)

Tie them where they intersect.
Amarrarlas donde intersecten.
ah-mah-RRRAHR-lahs DOHN-deh een-tehr-SEHK-tehn

Rebar location

It takes (eleven) bars.
Se necesitan (once) barras.
seh neh-seh-SEE-tahn (OHN-seh) BAH-rrrahs

...horizontal / vertical / in the corner.
...horizontal / vertical / en la esquina.
...oh-ree-sohn-TAHL / behr-tee-KAHL / ehn lah eh-SKEE-nah

Every (six) (feet / inches).
Cada (seis) (pies / pulgadas).
KAH-dah (sais) (pyehs / pool-GAH-dahs)

Center it / them in the forms.
Centrarla(s) en los moldes.
sehn-TRAHR-lah(s) ehn ehl MOHL-dehs

Let the rebar stick up (eight) (feet / inches).
Que la barra de refuerzo sobresalga (ocho) (pies / pulgadas).
keh lah BAH-rrrah deh reh-FOO-air-zoh soh-breh-SAHL-gah
(OH-choh) (pyehs / pool-GAH-dahs)

Figure 4. Overlapping Rebar

Cómo traslapar las barras de refuerzo

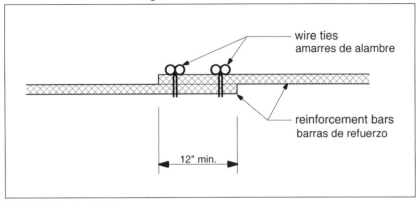

Overlap rebar at least 12 inches.

Bending rebar	**Bend the rebar here.** Doblar la barra de refuerzo aquí. *doh-BLAHR lah BAH-rrrah-deh rreh-FWEHR-soh ah-KEE* **Use the rebar bender.** Usar el doblador. *oo-SAHR ehl doh-blah-DOHR*
Cutting rebar	**Cut the rebar to (seventy nine) (feet / inches).** Cortar la barra de refuerzo a (setenta y nueve) (pies / pulgadas). *kohr-TAHR lah BAH-rrrah deh rreh-FWER-soh ah (seh-TEHN-tah ee NWEH-beh) (pyehs / pool-GAH-dahs)*
Tying rebar	**Tie the rebar.** Amarrar la barra. *ah-mah-RRRAHR lah BAH-rrrah* **Use the tool.** Usar la herramienta. *oo-SAHR lah eh-rrrah-MYEHN-tah* **Use two ties.** Usar dos ataduras. *oo-SAHR dohs ah-tah-DOO-rahs*
Bar overlaps	**Overlap (twenty-four) times the bar diameter.** Traslapar (veinte y cuatro) veces el diámetro de la barra. *trahs-lah-PAHR (BAIN-teh ee KWAH-troh) beh-SEHS ehl DYAH-meh-troh deh lah BAH-rrrah* **For this size rebar overlap (twenty-four) inches.** Para una barra de refuerzo de este tamaño, traslapar (veinte y cuatro) pulgadas. *PAH-rah OO-nah BAH-rrrah deh rreh-FWER-soh deh EH-steh tah-MAHN-yoh, trahs-lah-PAHR (BAIN-teh ee KWAH-troh) pool-GAH-dahs*
Supporting rebar	**It must be (two) inches off the ground.** Debe estar a (dos) pulgadas del piso. *DEH-beh eh-STAHR (dohs) pool-GAH-dahs dehl PEE-soh.*

Support it with wire.
Apoyarla con cables.
ah-poh-YAHR-lah kohn KAH-blehs

Put (supports / rocks) under it.
Poner (soportes / rocas) por debajo.
poh-NEHR (soh-POHR-tehs / RRROH-kahs)
pohr deh-BAH-hoh

Columns

Make a (rectangular / round) column.
Hacer una columna (rectangular / redonda).
ah-SEHR OO-nah koh-LOOM-nah
(rrrehk-tahn-goo-LAHR / reh-DOHN-dah)

Make it (six) inches by (twelve) inches.
Que tenga (seis) por (doce) pulgadas.
keh TEHN-gah (sais) pohr (DOH-seh) pool-GAH-dahs

Make it (sixteen) inches in diameter.
Que tenga (dieciséis) pulgadas de diámetro.
keh TEHN-gah (dyehs-ee-SAIS
pool-GAH-dahs deh DYAH-meh-troh

(Horizontal / Vertical) bars every (six) inches.
Barras (horizontales / Verticales) cada (seis) pulgadas.
BAH-rrras (oh-ree-sohn-TAHL-ehs / behr-tee-KAHL-ehs)
KAH-dah (sais) pool-GAH-dahs

▬▬▬▬▬▬▬▬ MIXING CONCRETE

Mixing concrete

Mix the concrete.
Mezclar el concreto.
mehs-KLAHR ehl kohn-KREH-toh

Put in (two) bags of (concrete / cement).
Poner (dos) bolsas de (concreto / cemento).
poh-NEHR (dohs) BOHL-sahs deh
(kohn-KREH-toh / seh-MEHN-toh)

(Twenty) shovels of sand per bag.
(Veinte) paladas de arena por bolsa.
(BAIN-teh) pah-LAH-dahs deh ah-REH-nah pohr BOHL-sah

(Two) buckets of water per bag.
(Dos) baldes de agua por bolsa.
(dohs) BAHL-dehs deh AH-gwah pohr BOHL-sah

Clean up

Clean (the wheelbarrow / the mixer).
Limpiar (la carretilla / el mezclador).
leem-PYAHR (lah kah-rreh-TEE-yah / ehl MEHS-klah-dohr)

...the shovel / the tools
...la pala / las herramientas
...lah PAH-lah / lahs eh-rrrah-MYEHN-tahs

■ POURING CONCRETE

Preparation

Get ready to pour concrete.
Prepararse para vaciar el concreto.
preh-pahr-AHR-seh PAH-rah bah-SYAHR ehl kohn-KREH-toh

The truck will be here in (twenty) (minutes / hours).
El camión llegará en (veinte) (minutos / horas).
ehl kahm-YOHN yeh-GAH-rah ehn (BAIN-teh)
(mee-NOO-tohs / OH-rahs)

Soaking forms

Soak (the forms / the sand / the soil).
Mojar (los moldes / la arena / la tierra).
moh-HAHR (lohs MOHL-dehs / lah
ah-REH-nah / la TYEH-rah)

Use the hose.
Usar la manguera.
oo-SAHR lah mahn-GEH-rah

Directing traffic

Direct traffic for the concrete truck.
Dirigir el tráfico para el camión hormigonero.
dee-ree-HEER ehl TRAH-fee-koh PAH-rah ehl
kahm-YOHN or-mee-goh-NEH-roh

Safety gear

Wear the orange vest.
Usar el chaleco anaranjado.
oo-SAHR ehl chah-LEH-koh ah-nah-rahn-HAH-doh

Use the sign.
Usar la señal.
oo-SAHR lah sehn-YAHL

Stand where cars can see you.
Quédese parado dónde lo vean los carros.
keh-deh-seh pah-RAH-doh DOHN-deh
loh beh-AHN lohs CAHR-rohs

Concrete truck

Clear a path for the concrete truck.
Quite todo del camino del camión hormigonero.
KEE-teh TOH-doh dehl kah-MEE-noh dehl
kahm-YOHN or-mee-goh-NEH-roh

Back the truck in.
Reversar el camión.
rrreh-behr-SAHR ehl kahm-YOHN

Start at the (front / back / right / left side).
Comenzar (adelante / atrás / a la derecha / a la izquierda).
koh-mehn-SAHR (ah-deh-LAHN-teh / ah-TRAHS /
ah lah deh-REH-chah / ah lah ees-KYEHR-dah)

Wheelbarrow

Use the wheelbarrow.
Usar la carretilla.
oo-SAHR la keh-rrreh-TEE-yah

Clear a path for the wheelbarrow.
Despejar una vía para la carretilla.
dehs-peh-HAR OO-nah BEE-ah
PAH-rah lah keh-rrreh-TEE-yah

Lay a board across it.
Poner una tabla encima.
poh-NEHR OO-nah TAH-blah ehn-SEE-mah

Chuting concrete

…more / stop / wait a minute
…más / pare / espere un minuto
…mahs / PAH-reh / ehs-PEHR-eh oon mee-NOO-toh

Put more (here / there).
Poner más (aquí / allá).
poh-NEHR mahs (ah-KEE / ah-YAH)

...a lot more / a little more
...mucho más / un poco más
...MOO-choh mahs / oon POH-koh mahs

Pumping concrete (snorkel)

You handle the tube.
Maneje el tubo.
mah-NEH-heh ehl TOO-boh

Pour in layers of (one / two) feet.
Vaciar capas de (un / dos) pie(s).
bah-SYAHR KAH-pahs deh (oon / dohs) pyehs

Keep moving.
Moverse siempre.
moh-BEHR-seh SYEHM-preh

Tamping concrete

Tamp the concrete.
Apisonar el concreto.
ah-pee-soh-NAHR ehl kohn-KREH-toh

Use the tamper.
Usar el apisonador.
oo-SAHR ehl ah-pee-sohn-ah-DOHR

Screeding concrete

Screed the concrete.
Enrasar el concreto.
ehn-rah-sahr ehl kohn-KREH-toh

Vibrating concrete

Vibrate the concrete.
Vibrar el concreto.
bee-BRAHR ehl kohn-KREH-toh

...not too much / that's enough
...no mucho / está bien
...noh MOO-choh / eh-STAH byehn

Tap the forms with your hammer.
Apisonar los moldes con el martillo.
ah-pee-soh-NAHR lohs MOHL-dehs kohn ehl mahr-TEE-yoh

Isolation joints	**Install the isolation joints.** Instalar las juntas de aislamiento. *een-stah-LAHR lahs HOON-tahs* *deh ah-ees-lah-MYEHN-toh*
	Put them around the edge. Ponerlas alrededor del borde. *poh-NEHR-lahs ahl-reh-deh-DOHR dehl BOHR-deh*
	Put it here. Ponerla aquí. *poh-NEHR-lah ah-KEE*
Embedded hardware	**Lay out (the bolts / the straps).** Colocar (los pernos / las bandas). *koh-loh-KARH (lohs PEHR-nohs / lahs BAHN-dahs)*
	Put (a bolt / a strap) here. Poner (un perno / una banda) aquí. *poh-NEHR (oon PEHR-noh / OO-nah BAHN-dah) ah-KEE*
Anchor bolts	**Put one every (six) feet.** Poner uno cada (seis) pies. *poh-NEHR OO-noh KAH-dah (sais) pyehs*
	Put them (two) inches in from the edge. Ponerlos a (dos) pulgadas del borde. *poh-NEHR-lohs (dohs) pool-GAH-dahs dehl BOHR-deh*
	Put them (two / and one-half) inches in from the edge. Ponerlos a (dos / y media) pulgadas del borde. *poh-NEHR-lohs (dohs / ee MEH-dyah)* *pool-GAH-dahs dehl BOHR-deh*
Clean up	**Clean (the wheelbarrow / the mixer).** Limpiar (la carretilla / el mezclador). *leem-PYAHR (lah kah-rrreh-TEE-yah / ehl MEHS-klah-dohr)*
	…the shovel / the tools …la pala / las herramientas *…lah PAH-lah / lahs eh-rrrah-MYEHN-tahs*

FINISHING CONCRETE

Finishing concrete

Finish the concrete.
Acabar el concreto.
ah-kah-BAHR ehl kahn-KREH-toh

Do you know how to finish concrete?
¿Sabe acabar el concreto?
SAH-beh ah-koh-BAHR ehl kohn-KREH-toh

Watch, do it like this.
Mire, así.
MEE-reh, ah-SEE

Floating concrete

Float the concrete.
Flotar el concreto.
floh-TAHR ehl kohn-KREH-toh

Watch out for the power line.
Cuidado con el cable de energía.
kwee-DAH-doh kohn ehl KAH-bleh deh eh-nehr-HEE-yah

Work the other direction, too.
Trabajar en el otro sentido también.
*trah-bah-HAHR ehn ehl OH-troh
sehn-TEE-doh tahm-BYEHN*

Troweling concrete

Trowel the concrete.
Paletear el concreto.
pah-leh-teh-AHR ehl kohn-KREH-toh

Wait until the water is gone.
Esperar que se vaya el agua.
ehs-peh-RAHR keh seh BAH-yah ehl AH-gwah

Kneel on a board.
Arrodillarse sobre una tabla.
ah-rrroh-dee-YAHR-seh SOH-breh OO-nah TAH-blah

Power trowel

Do you know how to operate a power trowel?
¿Sabe operar una paleta eléctrica?
*SAH-beh oh-peh-RAHR OO-nah
pah-LEH-tah eh-LEHK-tree-kah*

Control joints	**Cut in the control joints.** Cortar las juntas de control. *kohr-TAHR las hoon-tahs deh kohn-TROHL* **Put one every (ten / twelve) feet.** Poner una cada (diez / doce) pies. *poh-NEHR KAH-dah (dyehs / DOH-seh) pyehs* **Make them (one and one-quarter) inches deep.** Que tenga (una y un cuarto de) pulgada(s) de profundidad. *keh TEHN-gah (OO-nah ee oon KWAHR-toh deh)* *pool-GAH-dahs deh proh-foon-dee-DAHD*
Edging concrete	**Edge the concrete.** Bordear el concreto. *bohr-deh-AHR ehl kohn-KREH-toh*
Brooming concrete	**Broom (the walk / the driveway).** Barrer (la acera / la entrada). *bah-RRREHR (lah ah-SEHR-ah / lah ehn-TRAH-dah)* **Watch, do it like this.** Mire, así. *MEE-reh ah-SEE*
Drying time	**Wait (fifteen) minutes.** Esperar (quince) minutos. *ehs-pehr-AHR (KEEN-seh) mee-NOO-tohs*
Dying concrete	**(Throw on / add) the dye.** (Echar / añadir) la tintura. *(eh-CHAR / ahn-yah-DEER) lah teen-TOO-rah* **Use (red / black).** Usar (rojo / negro). *OO-sar (RRROH-hoh / NEH-groh)* **…not too much / a little more** …no mucho / un poco más *…noh MOO-choh / oon POH-koh mahs*

Stamp the concrete.
Hollar el concreto.
oh-LAHR ehl kohn-KREH- toh

Use the (stone / tile) stamp.
Usar el hollador de (piedra / baldosa).
oo-SAHR ehl oh-lah-DOHR deh (PYEH-drah / bahl-DOH-sah)

Do it carefully.
Hacerlo con cuidado.
ah-SEHR-loh kohn kee-DAH-doh

CURING CONCRETE

Roll out the plastic.
Extender el plástico.
ehk-steh-DEHR ehl PLAH-stee-koh

Remove the wrinkles.
Sacar las arrugas.
sah-KAR lahs ah-RRROO-gahs

Tape the edges.
Encintar los bordes.
ehn-seen-TAHR lohs BOHR-dehs

Spray it.
Rociarlo.
roh-SYAHR-loh

Wait until the water is gone.
Esperar que se vaya el agua.
ehs-peh-RAHR keh seh BAH-yah ehl AH-gwah

Nail two by four on top of the forms.
Clavar dos por cuatro sobre los moldes.
klah-BAHR dohs pohr KWAH-troh SOH-breh lohs MOHL-dehs

Caulk the joints.
Calafatear las juntas.
kah-lah-fah-teh-AHR lahs hoon-tahs

Fill it with (two) inches of water.
Llenarlo con (dos) pulgadas de agua.
yeh-NAHR-loh kohn (dohs) pool-GAH-dahs deh AH-gwah

Freeze protection

Cover the concrete.
Cubrir el concreto.
koo-BREER ehl kohn-KREH-toh

Use (the blankets / straw and plastic).
Usar (las sábanas / paja y plástico).
oo-SAHR (lahs sah-BAH-nahs / PAH-hah ee PLAH-stee-koh)

■ BACKFILL AND COMPACTION

Backfilling

Backfill around the foundation.
Llenar alrededor de la fundación.
Yay-NAHR ahl-reh-deh-DOHR deh lah foon-dah-SYOHN

Put in (six) inches at a time.
Poner (seis) pulgadas a la vez.
poh-NEHR (sais) pool-GAH-dahs ah lah behs

Compacting soil

Compact the soil.
Compactar la tierra.
kohm-pahk-TOHR lah TYEH-rah

Do you know how to operate the tamper?
¿Sabe operar el apisonador?
SAH-beh oh-pehr-RAHR el ah-pee-sohn-ah-DOHR

Do you need gas?
¿Necesita gas?
neh-seh-SEE-tahs gas

SETTING UP THE CUT YARD

Setting up the radial arm saw location

Set up the saw (here / there).
Montar la sierra (aquí / allá).
mohn-TAR lah SYEHR-ah (ah-KEE / ah-YAH)

Leveling the saw

It must be level.
Debe estar nivelada.
DEH-beh eh-STAHR-nee-beh-LAH-dah

Slope it back a little.
Inclinarla un poco hacia atrás.
een-klee-NAHR-lah oon POH-koh AH-syah ah-TRAHS

Run-off table

Make a run-off table.
Armar una mesa de apoyo.
ahr-MAHR OO-nah MEH-sah deh ah-POH-yoh

Use two by twelve.
Usar dos por doce.
oo-SAHR dohs pohr DOH-seh

Nail a two by four to the back.
Clavar un dos por cuatro atrás.
klah-BAHR oon dohs por DOH-seh ah-TRAHS

Support the ends.
Apoyar las puntas.
ah-poh-YAHR lahs POON-tahs

CUTTING HEADER AND CRIPPLE LOADS

Using stops

Use a stop.
Usar un tope.
oo-SAHR oon TOH-peh

Measure (thirty) inches from the blade.
Medir (treinta) pulgadas desde la cuchilla.
Meh-DEER (TRAIN-tah) pool-GAH-dahs

Mark the table.
Marcar la mesa.
mahr-KAHR lah MEH-sah

Nail the block.
Clavar el bloque.
klah-BAHR ehl BLOH-keh

Gang cutting　　**Cut (six) at a time.**
Cortar (seis) al mismo tiempo.
kohr-TAHR (sais) ahl MEES-moh TYEHM-poh

They must touch the stop.
Deben tocar el tope.
DEH-behn toh-KAHR ehl TOH-peh

Cut list　　**Cut everything on this list.**
Cortar todo en esta lista.
kohr-TAHR TOH-doh ehn EH-stah LEE-stah

Cut these first.
Cortar éstos primero.
kohr-TAHR EH-stohs pree-MEH-roh

Cut (fifty) more of these.
Cortar (cincuenta) más de éstos.
kohr-TAHR (seen-KWEHN-tah) mahs deh EH-stohs

Using up scrap　　**Cut the long ones first.**
Cortar los largos primero.
kohr-TAHR lohs LAHR-gohs pree-MEH-roh

Cut the shorter ones from the scrap.
Cortar los más cortos utilizando los desperdicios.
kohr-TAHR lohs mahs KOHR-tohs
oo-tee-lee-SAHN-doh lohs dehs-pehr-DEE-syohs

Building loads　　**Put stickers down.**
Poner los soportes por debajo.
poh-NEHR lohs soh-POHR-tehs pohr deh-BAH-hoh

Put the long ones on the bottom.
Poner los largos abajo.
poh-NEHR lohs LAHR-gohs ah-BAH-hoh

Put the short ones on top.
Poner los cortos arriba.
poh-NEHR lohs KOHR-tohs ah-RRREE-bah

Be careful.
Cuidado.
kwee-DAH-doh

Don't put your hand in front of the blade.
No ponga su mano frente a la cuchilla.
*noh POHN-gah soo MAH-noh FREHN-teh
lah koo-CHEE-yah*

If the saw binds it will kick.
Si la sierra se atranca, se sacudirá.
*see lah SYEH-rrrah seh ah-TRAHN-kah,
seh sah-koo-DEE-rah*

When you aren't using it, turn it off.
Cuando no esté en uso, apagarla.
KWAHN-doh noh EH-steh ehn OO-soh, ah-pah-GAHR-lah

FRAMING WALLS

Clear the floor.
Despejar el piso.
dehs-peh-HAHR ehl PEE-soh

Move those walls.
Mover esas paredes.
moh-BEHR EH-sahs pah-REH-dehs

Mark numbers on the walls and floor.
Marcar números en las paredes y el piso.
*mahr-KAHR NOO-meh-rohs eh lah
pah-REH-dehs ee ehl PEE-soh*

Figure 5. Wall Framing Members
Elementos estructurales de la pared

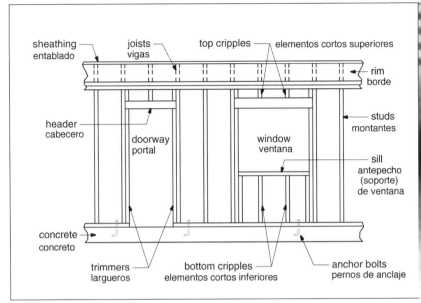

Framing members in a typical stud-framed wall.

Move them straight out.
Sacarlas.
sah-KAHR-lahs

Laying out walls

Frame this wall here.
Construir esta pared aquí.
kohn-stroo-EER EH-stah pah-REHD ah-KEE

Frame this wall on top of that one.
Construir esta pared sobre aquélla.
kohn-stroo-EER EH-stah pah-REHD SOH-breh ah-KEH-yah

Laying out studs

Laying out the studs.
Colocar los montantes.
koh-loh-KARH lohs mohn-TAHN-tehs

Following layout

One mark; one stud.
Una marca; un montante.
OO-nah MAHR-kah, OO-nah mohn-TAHN-teh

Put it on this side of the line.
Ponerlo en este lado de la línea.
poh-NEHR-loh ehn EH-steh LAH-doh deh lah LEEN-yah

That stud goes flat.
Ese montante va plano.
EH-seh mohn-TAHN-teh bah PLAH-noh

Crowning studs

Crown the studs.
Coronar los montantes.
koh-roh-NAHR lohs mohn-TAHN-tehs

Sight the edge.
Visar el borde.
bee-SAHR ehl BOHR-deh

The top of the curve is the crown.
El tope de la curva es la corona.
ehl TOH-peh deh lah KOOR-bah ehs lah koh-ROH-nah

Put the crowns (up / down).
Poner las coronas hacia (arriba / abajo).
poh-NEHR lahs koh-ROH-nahs AH-syah
(ah-RRREE-bah / ah-BAH-hoh)

Separating wall plates

Pull the plates apart.
Separar las placas.
seh-pah-RAHR lahs PLAH-kahs

Put the top plate here.
Poner la placa superior aquí.
poh-NEHR lah PLAH-kah soo-pehr-YOHR ah-kee

Keep the layout marks up.
Mantener las marcas del trazado hacia arriba.
mahn-teh-NEHR lahs MAHR-kahs dehl
trah-zah-doh AH-syah ah-RRREE-bah

That plate is backwards.
Esa placa está al revés.
EH-sah plah-KAH EH-stah ahl rrreh-BEHS

Nailing studs

Nail the studs.
Clavar los montantes.
klah-BAHR lohs mohn-TAHN-tehs

Keeping the wall square

Keep the wall square.
Mantener la pared cuadrada.
mahn-teh-NEHR lah peh-REHD kwah-DRAH-dah

Nailing schedule

(Two / three) nails per stud.
(Dos / tres) clavos por montante.
(dohs / trehs) KLAH-bohs pohr mohn-TAHN-teh

Placing flat studs

That stud goes (up / down).
Ese montante va (arriba / abajo).
EH-seh mohn-TAHN-teh bah
(ah-RRREE-bah / ah_BAH-hoh)

Laying out top plate

Laying out the top plate.
Colocar la placa superior.
koh-loh-KARH lah PLAH-kah soo-pehr-YOHR

Overlap the plates.
Traslapar las placas.
trahs-lah-PAHR lah PLAH-kahs

Extend it past (three / five) and one-quarter inches.
Extenderla más de (tres / cinco) y un cuarto de pulgada(s).
ehk-stehn-DEHR-lah mahs deh (trehs / SEEN-koh
ee oon KWAHR-toh deh pool-GAH-dah(s)

Hold it back from the line (one quarter-inch).
Apartarla de la línea (un cuarto) de pulgada.
ah-pahr-TAHR-lah deh lah LEEN-yah
(oon KWAHR-toh) deh pool-GAH-dah

Cut the top plate.
Cortar la placa superior.
kohr-TAHR lah PLAH-kah soo-pehr-YOHR

Cut it here.
Cortarla aquí.
kohr-TAHR-lah ah-KEE

Nail the top plate.
Clavar la placa superior.
klah-BAHR lah PLAH-kah soo-pehr-YOHR

One nail every (sixteen) inches, staggered.
Un clavo cada (dieciséis) pulgadas, alternándose.
oon KLAH-boh KAH-dah (dyehs-ee-SAIS)
pool-GAH-dahs ahl-terhn-ahn-doh-seh

Two nails at (the break / the end).
Dos clavos en (punto de unión / la punta).
dohs KLAH-bohs ehn (POON-toh deh oo-NEE-ohn /
la POON-tah)

Leave this open for the beam.
Dejar abierto aquí para la viga.
deh-HAHR ah-BYEHR-toh ah-KEE PAH-rah lah BEE-gah

Make it (four / six) inches wide.
Que tenga (cuatro / seis) pulgadas de ancho.
keh TEHN-gah (KWAH-troh / sais)
pool-GAH-dahs deh AHN-choh

Cut out the top plate.
Cortar la placa superior.
kohr-TAHR lah PLAH-kah soo-pehr-YOHR

Snapping block lines | **Snap a line for blocks.**
Marcar una línea con tiza para los bloques.
Marh-KAR OO-nah LEEN-yah kohn
TEE-zah PAH-rah lohs BLOH-kehs

(Ninety-six) inches off the floor.
A (noventa y seis) pulgadas del piso.
ah (noh-BEHN-tah ee sais) pool-GAH-dahs dehl PEE-soh

The bottom plate must be straight.
La placa inferior debe estar derecha.
lah PLAH-kah een-fehr-YOHR
DEH-beh eh-STAHR deh-REH-chah

Watch, do it like this.
Mire, así.
MEE-reh, ah-SEE

Cutting-in blocks | **Cut the blocks.**
Cortar los bloques.
kohr-TAHR lohs BLOH-kehs

Use scrap.
Usar los desperdicios.
oo-SAHR lohs dehs-pehr-DEE-syohs

Watch, do it like this.
Mire, así.
MEE-reh, ah-SEE

Cutting let-in braces | **Cut in the let-in braces.**
Cortar las riostras de entrada.
kohr-TAHR lohs rrree-OH-strahs deh ehn-TRAH-dah

Figure 6. Let-In Brace
Riostras de entrada

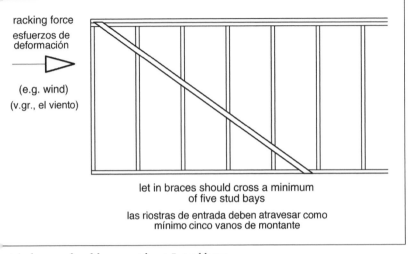

racking force
esfuerzos de
deformación

(e.g. wind)
(v.gr., el viento)

let in braces should cross a minimum
of five stud bays

las riostras de entrada deben atravesar como
mínimo cinco vanos de montante

Let-in braces should cover at least 5 stud bays.

It must cover five bays.
Debe cubrir cinco vanos.
DEH-beh KOO-breer SEEN-koh BAHN-ohs

Nail the bottom only.
Clavar sólo la parte inferior.
klah-BAHR SOH-loh lah PAHR-teh een-fehr-YOHR

Tack the nails, three per stud.
Fijar los clavos, tres por montante.
fee-HAHR lohs KLAH-bohs, trehs pohr mohn-TAHN-teh

Laying out temporary braces

Lay out braces for the walls.
Colocar las riostras para las paredes.
koh-loh-KARH lahs rrree-OH-strahs
PAH-rah lahs pah-REH-dehs

Put one (at each end / in the middle).
Poner una (en cada punta / en el medio).
poh-NEHR OO-nah (ehn KAH-dah
POON-tah / ehn ehl MEH-dyoh)

Standing walls

Stand the wall.
Levantar la pared.
leh-bahn-TAHR lah pah-REHD

We need (two / three / four) men.
Necesitamos (dos / tres / cuatro) hombres.
neh-seh-see TAH-mohs (dohs / trehs /
KWAH-troh) OHM-brehs

This wall (first / second).
Esta pared (primero / segundo).
EH-stah pah-REHD (pree-MEH-roh / seh-GOON-doh)

Hold the wall.
Sostener la pared.
soh-steh-NEHR lah pah-REHD

Temporary brace locations

Brace the wall.
Arriostrar la pared.
rrree-oh-STRAHR lah pah-REHD

Put one here.
Poner una aquí.
poh-NEHR OO-nah ah-KEE

Put one (at each end / in the middle).
Poner una (en cada punta / en el medio).
*poh-NEHR OO-nah (ehn KAH-dah
POON-tah / ehn ehl MEH-dyoh)*

Bracing walls

Nail the (top / bottom) of the brace.
Clavar la parte (superior / inferior) de la riostra.
*klah-BAHR lah PAHR-teh (soo-pehr-YOHR /
een-fehr-YOHR) deh lah rrree-OH-strah*

Lean the top out more.
Inclinar el tope hacia afuera más.
een-klah-NAHR ehl TOH-peh AH-syah ah-FWEH-rah, mahs

Plumb it.
Aplomarla.
ap-loh-MARH-la

Nail it.
Clavarla.
klah-BAHR-lah

Nailing off corners

Nail off the corners.
Clavar las esquinas.
klah-BAHR lahs eh-SKEE-nahs

(Two) nails every (twenty-four) inches.
(Dos) clavos cada (veinticuatro) pulgadas.
*(dohs) KLAH-bohs KAH-dah
(BEHN-tee KWAH-troh) pool-GAH-dahs*

Stagger the nails.
Alternar la colocación de los clavos.
ahl-terh-NAR lah koh-LO-eh-sien deh lohs KLAH-bohs

It must be flush.
Debe quedar a ras.
DEH-beh keh-DAHR ah rrrahs

Nailing off top plate	**Nail off the top plate.** Clavar la placa superior. *klah-BAHR lah PLAH-kah soo-pehr-YOHR* **Put (two) nails through the corner.** Poner (dos) clavos a través de la esquina. *poh-NEHR (dohs) KLAH-bohs ah trah-BEHS* *deh lah eh-SKEE-nah* **Put a toenail here.** Poner un clavo sesgado aquí. *poh-NEHR oon KLAH-boh sehs-GAH-doh ah-KEE*

◼ PLUMB AND LINE

Nutting down	**Put the (nuts / washers) on the bolts.** Poner las (tuercas / arandelas) en los pernos. *poh-NEHR las (TWEHR-kahs / ah-rahn-DEH-lahs)* *ehn lohs PER-nohs* **Tighten down the nuts.** Apretar las tuercas. *ah-preh-TAHR lahs TWEHR-kahs* **It must be on the line.** Debe quedar en la línea. *DEH-beh keh-DAHR ehn lah LEEN-yah*
Shooting down	**Shoot down the plates.** Rociar las placas. *rrroh-SYAHR lahs PLAH-kahs* **Use safety glasses.** Usar gafas de seguridad. *oo-SAHR GAH-fahs deh seh-goo-ree-DAHD* **It must be on the line.** Debe quedar en la línea. *DEH-beh keh-DAHR ehn lah LEEN-yah*

in locations	**Put one here.** Poner uno aquí. *poh-NEHR OO-noh ah-KEE*
aying out braces	**Laying out the braces.** Colocar las riostras. *koh-loh-KARH lahs rrree-OH-strahs*
	One brace per wall. Una riostra por pared. *OO-nah rrree-OH-strah pohr pah-REHD*
ailing braces	**Nail the braces.** Clavar las riostras. *klah-BAHR lahs rrree-OH-strahs*
	Nail the (tops / the bottom) only. Clavar sólo la parte (superior / inferior). *klah-BAHR SOH-loh lah PAHR-teh* *(soo-pehr-YOHR / een-fehr-YOHR)*
	Put it (in the middle / at the break). Ponerlo en el (medio / punto de unión de los extremos). *poh-NEHR-loh ehn ehl (MEH-dyoh /* *OON-toh deh oo-NEE-ohn)*
lumbing walls	**Plumb the walls.** Aplomar las paredes. *ap-loh-MARH lahs pah-REH-dehs*
	Plumb the outside corners first. Aplomar las esquinas exteriores primero. *ap-loh-MARH lahs eh-SKEE-nahs* *ehk-stehr-YOHR-ehs pree-MEH-roh*
	It must (come in / go out). Debe (entrar / salir). *DEH-beh (ehn-TRAHR / sah-LEER)*
	(That's too much / more). Nail it. *(Demasiado / más). Clavarla.* *(deh-mah-SYAH-doh / mahs) klah-BAHR-lah*

Racking walls	**Rack the wall.** Empujar la pared. *ehm-poo-HAHR lah pah-REHD*
	It must (come to me / go to you). Debe (venir hacia mí / ir hacia usted). *DEH-beh (beh-NEER AH-syah mee /* *eer AH-syah oo-STEHD)*
Aligning walls	**Align the walls.** Alinear las paredes. *ah-leen-YAHR lahs pah-REH-dehs*

▮ SHEAR PANEL

Shear location	**Put plywood from here to here.** Poner madera terciada de aquí a aquí. *poh-NEHR mah-derh-ah terh-SEE-eh-da* *deh ah-KEE ah ah-KEE*
	Put plywood on the whole exterior. Poner madera terciada en todo el exterior. *poh-NEHR mah-derh-ah terh-SEE-eh-da* *ehn TOH-doh ehl ehk-stehr-YOHR*
	Put plywood on this (wall / side). Poner madera terciada en (esta pared / este lado). *poh-NEHR mah-derh-ah terh-SEE-eh-da ehn* *(EH-stah pah-REHD / EH-steh LAH-doh)*
Method of **installation**	**Run it (horizontal / vertical).** Instalarlo (horizontal / vertical). *een-stah-LAHR-loh (oh-ree-sohn-TAHL / behr-tee-KAHL)*
	Cut it before you install it. Cortarlo antes de instalarlo. *kohr-TAHR-loh AHN-tehs deh een-stah-LAHR-loh*
	Install it over the openings, then cut them out. Instalarlo sobre las aberturas, y luego cortarlas. *een-stah-LAHR-loh SOH-breh lahs* *ah-behr-TOO-rahs, ee LWEH-go kohr-TAHR-lahs*

dge blocking	**Block the edges.** Bloquear los bordes. *bloh-keh-AHR lohs BOHR-dehs* **Watch, do it like this.** Mire, así. *MEE-reh, ah-SEE*
ying corners together	**Extend it past the end of the wall.** Extenderlo más allá del final de la pared. *ehk-stehn-DEHR-loh mahs ah-YAH* *dehl fee-NAHL deh lah pah-REHD* **Run it past (three / five) and a quarter inches.** Correrlo más de (tres / cinco) y un cuarto de pulgadas. *koh-RRREHR-loh mahs de (trehs / SEEN-koh)* *ee oon KWAHR-toh deh pool-GAH-dahs*
Nailing shear panel	**Nail the plywood.** Clavar la madera terciada. *klah-BAHR lah mah-derh-ah terh-SEE-eh-da* **Use the (nail / staple) gun.** Usar la (clavadora / engrapadora). *oo-SAHR lah (klah-bah-DOH-rah / koh-seh-DOH-rah)*
lush-nailing	**The nails must be flush.** Los clavos deben quedar parejos. *lohs KLAH-bohs DEH-behn keh-DAHR pah-REH-hohs* **Use the flush-nailer.** Usar la clavadora parejo. *oo-SAHR lah klah-bah-DOH-rah pah-REH-hoh* **Turn the pressure (up / down). More.** (Subir / bajar) la presión. Más. *(soo-BEER / bah-HAHR) lah preh-SYOHN. Mahs*
Nails	**Use (eights / staples).** Usar (ochos / ganchos). *oo-SAHR (OH-chohs / GAHN-chohs)*

Nailing schedules	**Nail it every (three / four / six) inches at the edges.**
	Clavarlo cada (tres / cuatro / seis) pulgadas en los bordes.
	klah-BAHR-loh KAH-dah (trehs / KWAH-troh / sais)
	pool-GAH-dahs ehn lohs BOHR-dehs
	Nail it every twelve inches in the middle.
	Clavarlo cada doce pulgadas en el medio.
	klah-BAHR-loh KAH-dah DOH-seh
	pool-GAH-dahs ehn ehl MEH-dyoh
	Don't nail the trimmers.
	No clavar los largueros.
	noh klah-BAHR lohs lahr-GEH-rohs

▬▬▬▬▬▬▬▬ BEAMS

Crowning beams	**Crown the beams.**
	Coronar las vigas.
	koh-roh-NAHR lahs BEE-gahs
	The crown must be up.
	La corona debe estar hacia arriba.
	lah koh-ROH-nah DEH-beh
	eh-STAHR AH-syah ah-RRREE-bah
Laying out beams	**Lay out the beams.**
	Colocar las vigas.
	koh-loh-KARH lahs BEE-gahs
	The locations are marked on the floor.
	Las ubicaciones están marcadas en el piso.
	lahs oo-bee-kah-SYOHN-ehs EH-stahn
	mahr-KAH-dahs ehn ehl PEE-soh
	Get some help.
	Conseguir ayuda.
	kohn-seh-GEER ah-YOO-dah
Setting beams	**Set the beams.**
	Instalar las vigas.
	ihn-stahl-arh lahs BEE-gahs

Be careful!
¡Cuidado!
kwee-DAH-doh

Put it up here.
Levantarla aquí.
leh-bahn-TAHR-lah ah-KEE

Beam hardware | **Put a (strap / clip) here.**
Poner (una correa / un clip) aquí.
poh-NEHR (OO-nah koh-RRREH-ah / oon clip) ah-KEE

JOISTS

Joist type | **This takes (TJI / two by six) joists.**
Esto toma vigas de (TJI / dos por seis).
EH-stoh TOH-mah BEE-gahs deh (TJI / dohs pohr sais)

Joist location | **These joist are for (here / there).**
Estas vigas son para (aquí / allá).
EH-stahs BEE-gahs sohn PAH-rah (ah-KEE / ah-YAH)

The layout is marked on the floor.
La disposición está marcada en el piso.
lah dees-poh-see-SYOHN EH-stah
mahr-KAH-dah ehn ehl PEE-soh

Put a joist above every wall.
Poner una viga sobre cada pared.
poh-NEHR oon BEE-gah SOH-breh KAH-dah pah-REHD

Joist direction | **They run this way.**
Corren en esta dirección.
koh-RRREHN ehn EH-stah dee-rehk-SYOHN

They change direction here.
Cambian de dirección aquí.
kahm-BYAHN deh dee-rehk-SYOHN ah-KEE

Pulling joist layout | **(Butt / hook) the wall.**
(Tocar / enganchar) la pared.
(toh-KAHR / ehn-GAHN-char) lah pah-REHD

(Fifteen and one-quarter) inches to the first one.
A (quince y un cuarto) pulgadas de la primera.
ah (KEEN-seh ee oon KWAHR-toh)
pool-GAH-dahs deh lah pree-MEH-rah

Sixteen-inch centers **These go on (sixteen)-inch centers.**
Éstas van centradas a (dieciséis) pulgadas.
EH-stahs bahn sehn-TRAH-dahs ah
(dyehs-EE-sais) pool-GAH-dahs

Doublers / triplers **Put a (doubler / tripler) here.**
Poner uno (doble / triple) aquí.
poh-NEHR OO-noh (DOH-bleh / TREE-pleh) ah-KEE

Nail them together.
Clavarlos juntos.
klah-BAHR-lohs HOON-tohs

(Sixteen / twenty-four) inches on center.
A (dieciséis / veinticuatro) pulgadas centradas.
ah (dyehs-ee-SEHS / behn-tee-KWAH-troh)
pool-GAH-dahs sehn-TRAH-dahs

They must be flush.
Deben quedar a ras.
DEH-behn keh-DAHR ah rrrahs

Laying out joists **Lay out the joists.**
Extender las vigas.
ehk-sten-DER lahs BEE-gahs

Lay out this side (first / second).
Esparcir este lado (primero / segundo).
es-pahr-SEER EH-steh LAH-doh
(pree-MEH-roh / seh-GOON-doh)

Put them on the layout marks.
Ponerlas sobre las marcas del trazo.
poh-NEHR-lahs SOH-breh lahs MAHR-kahs dehl TRAH-zoh

Overlap them in the middle.
Traslaparlas en el medio.
trahs-lah-PAHR-lahs ehn ehl MEH-dyoh

Laying out joists (into hangers)

Put glue in the hanger.
Poner pegamento en el colgante.
poh-NEHR PEH-gah-mehn-toh ehn ehl kohl-GAHN-teh

It is too (long / short).
Es demasiado (larga / corta).
ehs deh-mah-SYAH-doh (LAHR-gah / KOHR-tah)

We must lower it at the same time.
Debemos bajarla al mismo tiempo.
deh-BEH-mohs bah-HAHR-lah ahl MEES-moh TYEHM-poh

Crowning joists

Crown the joists.
Coronar las vigas.
koh-roh-NAHR lahs BEE-gahs

The crowns must be up.
Las coronas deben quedar hacia arriba.
*lahs koh-ROH-nahs DEH-behn
keh-DAHR AH-syah ah-RRREE-bah*

Overhangs

There is an overhang here.
Hay un voladizo aquí.
AH-ee oon boh-lah-DEE-soh ah-KEE

It goes from here to here.
Va de aquí a aquí.
bah deh ah-KEE ah ah-KEE

Do not cut them off at the wall.
No cortarlas donde termina la pared.
*noh kohr-TAHR-lahs DOHN-deh
tehr-MEE-nah lah pah-REHD*

Laying out blocks

Lay out the (thirteen)-inch blocks first.
Colocar bloques de (trece) pulgadas primero.
*koh-loh-KARH BLOH-kehs deh (TREH-seh)
pool-GAH-dahs pree-MEH-roh*

Put one on each joist.
Poner uno en cada viga.
poh-NEHR OO-noh ehn KAH-dah BEE-gah

Lay out the (fourteen and one half)-inch blocks.
Colocar los bloques de (catorce y media) pulgadas.
koh-lo-KARH lohs BLOH-kehs deh
(kah-TOHR-seh ee MEH-dyah) pulgadas

Laying out rim

Lay out all the rim.
Colocar las vigas del perímetro.
koh-loh-KARH lahs-BEE-gahs-dehl

Use (this / that) for the rim.
Usar (ésta / ésa) para el perímetro.
oo-SAHR (eh-STAH / EH-seh) PAH-rah ehl

Double rim here to here.
Viga doble en el perímetro de aquí a aquí.
BEE-gah Doh-bleh ehn ehl

The rim stops here.
La viga del perímetro termina aquí.
Lah BEE-gah dehl tehr-MEE-nah ah-KEE

Nailing rim

Make it flush with the outside of the wall.
Que quede a ras con el exterior de la pared.
keh KEH-deh ah rrras-SAHN-teh kohn
ehl ehk-stehr-YOHR deh lah pah-REHD

One toenail every (sixteen) inches.
Un clavo sesgado cada (dieciséis) pulgadas.
oon KLAH-boh sehs-GAH-doh KAH-dah
(dyehs-ee-SAIS) pool-GAH-dahs

Let the ends run wild.
Dejar las puntas libres.
deh-HAHR lahs POON-tahs LEE-brehs

Cutting rim

Cut the rim at the corners.
Cortar la viga del perímetro en las esquinas.
kohr-TAHR lah BEE-gah dehl ehn lahs eh-SKEE-nahs.

Rolling joists

Roll the joists.
Levantar las vigas.
leh-bahn-TAHR lahs BEE-gahs

Start (here / in the middle).
Empezar (aquí / en el medio).
ehm-peh-SAHR (ah-KEE / ehn ehl MEH-dyoh)

Push it tight to the rim.
Empujarlo apretado hasta la viga del perímetro.
ehm-poo-HAHR-loh ah-preh-TAH-doh
AH-stah lah BEE-gah dehl

Nailing schedules **(Two) nails into the block.**
(Dos) clavos en el bloque.
(dohs) KLAH-bohs ehn ehl BLOH-keh

(Two / three / four) nails through the lap.
(Dos / tres / cuatro) clavos a través del traslapo.
(dohs / trehs / KWAH-troh) KLAH-bohs
ah trah-BEHS dehl trahs-LAH-poh

Toenail the joist to the plate.
Clavar sesgada la viga a la placa.
klah-BAHR sehs-GAH-dah lah BEE-gah ah lah PLAH-kah

Checking layout **Check the layout.**
Revisar las marcas de transferencia.
rrreh-bee-SAHR lahs deh trahns ferh ehn cyah

They must be within a (half) inch.
Deben estar dentro de (media) pulgada.
DEH-behn eh-STAHR dehn-TROH deh
(MEH-dyah) pool-GAH-dah

Checking for straightness **The joists must be straight.**
Las vigas deben estar derechas.
lahs BEE-gahs DEH-behn eh-STAHR deh-REH-chahs

Cutting special blocks **Cut a special block.**
Cortar un bloque especial.
kohr-TAHR oon BLOH-keh eh-speh-SYAHL

Measuring cut-lines **Mark the joists at the corners.**
Marcar las vigas en las esquinas.
mahr-KAHR las BEE-gahs ehn lahs eh-SKEE-nahs

Mark (one and one-half) inch(es) from the outside.
Marcar a (una y media) pulgada(s) del exterior.
mahr-KAHR ah (OO-nah ee MEH-dyah)
pool-GAH-dah(s) dehl ehk-stehr-YOHR

Marking cut-lines

Snap a line.
Marcar una línea con tiza.
marh-KAHR OO-nah LEEN-yah kohn TEE-zah

With your square, mark the joists.
Marcar las vigas con escuadrarse.
mahr-KAHR lahs BEE-gahs kohn ehs-kwah-DRARH-seh

Cutting joists

Cut the joists.
Cortar las vigas.
kohr-TAHR lahs BEE-gahs

A little short is better.
Corta es mejor.
KOHR-tah ehs meh-HOHR

The cuts must be square.
Los cortes deben escuadrarse.
lohs KOHR-tehs DEH-behn ehs-kwah-DRARH-seh

Marking overhangs (cantelevers)

Mark the overhang.
Marcar el voladizo.
mahr-KAHR ehl boh-lah-DEE-soh

The overhang is (forty-eight) inches.
El voladizo es de (cuarenta y ocho) pulgadas.
ehl boh-lah-DEE-soh ehs deh
(kwah-REHN-tah ee OH-choh) pool-GAH-dahs

**Mark it at (forty-six and a half)
inches from the outside wall.**
Marcarlo a (cuarenta y seis y media)
pulgadas de la pared exterior.
mahr-KAHR-loh ah (kwah-REHN-tah ee sais ee MEH-dyah)
pool-GAH-dahs deh lah pah-REHD ehk-stehr-YOHR

Snap a line.
Marcar una línea con tiza.
mahr-KAHR OO-nah LEEN-yah kohn TEE-zah

Cutting overhangs

Cut the overhang.
Cortar el voladizo.
kohr-TAHR ehl boh-lah-DEE-soh

The cuts must be square.
Los cortes deben escuadrarse.
lohs KOHR-tehs DEH-behn ehs-kwah-DRARH-seh

Blocking above walls

Block above each wall.
Bloquear sobre cada pared.
bloh-keh-AHR SOH-breh KAH-dah pah-REHD

Nailing joists to interior walls

Nail the joist to the plate.
Clavar la viga a la placa.
klah-BAHR lah BEE-gah ah lah PLAH-kah

The joists must be straight.
Las vigas deben quedar alineadas.
lahs BEE-gahs DEH-behn keh-DAHR deh-REH-chahs

Laying out backing

Lay out the backing.
Colocar el respaldo.
koh-loh-KARH ehl rrreh-SPAHL-doh

Watch, do it like this.
Mire, así.
MEE-reh, ah-SEE

Backing material

Use two by (four / six) for backing.
Usar dos por (cuatro / seis) para el respaldo.
ooa-SAHR dohs pohr (KWAH-troh / sais)
PAH-rah ehl rrreh-SPAHL-doh

Cutting backing

Cut in the backing.
Cortar el respaldo.
kohr-TAHR el rrreh-SPAHL-doh

It goes from wall to wall.
Va de pared a pared.
bah deh pah-REHD ah pah-REHD

Cut a block for here.
Cortar un bloque para aquí.
kohr-TAHR oon BLOH-keh PAH-rah ah-KEE

Hanging backing	**Hang the backing from the joist.** Colgar el respaldo de la viga. *kohl-GAHR ehl rrreh-SPAHL-doh deh lah BEE-gah*
Nailing backing	**Nail the backing from (above / the ladder).** Clavar el respaldo desde (arriba / la escalera). *klah-BAHR ehl rrreh-SPAHL-doh DEHS-deh* *(ah-RRREE-bah / lah eh-skah-LEH-rah)*
	Nail it on top of the wall. Clavarlo en el tope de la pared. *klah-BAHR-loh ehn ehl TOH-peh deh lah pah-REHD*
Backing blocks	**Put backing blocks here.** Poner bloques de respaldo aquí. *poh-NEHR BLOH-kehs deh rrreh-SPAHL-doh ah-KEE*

████████████ JOIST HANGERS

Hanger locations	**Put hangers here.** Poner los colgantes aquí. *poh-NEHR lohs kohl-GAHN-tehs ah-KEE*
Pulling hanger layout	**Mark the hanger layout.** Marcar la colocación de los colgantes. *mahr-KAHR la Deh lohs kohl-GAHN-tehs* *deh lohs kohl-GAHN-tehs*
	They go on (sixteen)-inch centers. Van centrados a (dieciséis) pulgadas. *bahn sehn-TRAH-dohs ah (dyehs-EE-sais) pool-GAH-dahs*

Pulling layout	**(Butt / hook) the (wall / joist).** (Tocar / enganchar) la (pared / viga). *(toh-KAHR / ehn-gahn-CHAHR) lah (pah-REHD / BEE-gah)*
	(Fifteen and one-quarter) inches to the first one. A (quince y un cuarto de) pulgada(s) del primero. *ah (KEEN-seh ee oon KWAHR-toh)* *pool-GAH-dah(s) dehl pree-MEH-roh*
Hanger backing **(engineered lumber)**	**Fill the web on (this / both) sides.** Llenar en el centro de las vigas en (este / ambos) lado(s). *yay-NAHR ehn ehl cehn-TROH deh lahs* *BEE-gahs ehn ee ehn (EH-steh / AHM-bohs) LAH-doh(s)*
Hanger nailing	**Nail the hangers.** Clavar los colgantes. *klah-BAHR lohs kohl-GAHN-tehs*
	Nail them (from the ladder / from above). Clavarlos desde (la escalera / arriba). *klah-BAHR-lohs DEHS-deh* *(lah eh-skah-LEH-rah / ah-RRREE-bah)*
Keeping hangers **flush**	**They must be flush on the bottom.** Deben quedar a ras abajo. *DEH-behn keh-DAHR ah rrrahs-SAHN-tehs ah-BAH-hoh*
Hanger nails	**Use (Teco / hanger nails).** Usar clavos (Teco / de colgante). *oo-SAHR KLAH-bohs(TEE-koh / deh kohl-GAHN-teh)*
	Use (sixteens). Usar los clavos (dieciséis). *oo-SAHR lohs KLAH-bohs (dyehs-EE-sais)*

FLOOR SHEATHING

Sheet the floor.
Entablar el piso.
ehn-tah-blarh ehl PEE-soh

Sheathing size	**Use (three-quarter / one and one-eighth) inch plywood.**
	Usar madera terciada de (tres cuartos de / una y un octavo de) pulgada.
	oo-SAHR mah-derh-ah terh-SEE-eh-dah deh (trehs KWAHR-tohs de / OO-nah ee oon ohk-TAH-boh deh) pool-GAH-dah
Sheathing type	**Use (plywood / OSB).**
	Usar (madera terciada / OSB).
	oo-SAHR (mah-derh-ah terh-SEE-eh-dah / OSB)
	Use tongue and groove.
	Usar una lengüeta y ranura.
	oo-SAHR OO-nah lehn-GWEH-tah ee rrrah-NOO-rah
Sheathing orientation	**It must run across the joists.**
	Debe correr sobre las vigas.
	DEH-beh koh-RRREHR SOH-breh lahs BEE-gahs
	It changes direction here.
	Cambia de dirección aquí.
	KAHM-byah deh dee-rehk-SYOHN ah-KEE
Tools	**We need (the sledge hammer / the glue / the chalk).**
	Necesitamos (el acotillo / el pegamento / la tiza).
	neh-seh-see-TAH-mohs (el ah-koh-TEE-yoh / ehl pegh- / lah TEE-sah)
	...the saw / a new blade / the nail gun / gun nails
	...la sierra / una nueva cuchilla / la clavadora / los clavos
	...lah SYEH-rah / OO-nah NWEH-bah koo-CHEE-yah / lah klah-bah-DOH-rah / lohs KLAH-bohs
	Use the flush-nailer.
	Usar la clavadora parejo.
	oo-SAHR lah klah-bah-DOH-rah pah-REH-hoh
Sheathing layout	**Start at this end.**
	Comenzar en esta punta.
	koh-mehn-SAHR en EH-stah POON-tah

Measure and mark (forty-eight and one-eighth) inches.
Medir y marcar (cuarenta y ocho y un octavo de) pulgada.
meh-DEER ee mahr-KAHR (kwah-REHN-tah ee
OH-choh ee oon ohk-TAH-boh deh) pool-GAH-dah

Snap a line.
Marcar una línea con tiza.
marh-KARH OO-nah LEEN-yah kohn TEE-zah

Laying out sheathing **Lay out the sheathing.**
Colocar el entablado.
koh-loh-KARH ehl ehn-tah-blah-doh

Start here.
Comenzar aquí.
koh-mehn-SAHR ah-KEE

Put the tongue on this side.
Poner la lengüeta en este lado.
poh-NEHR lah lehn-GWEH-tah ehn EH-steh LAH-doh

Staggering breaks **Stagger the breaks.**
Alternar los puntos de unión de los extremos.
ahl-terh-narh lohs deh oo-NEE-ohn deh lohs exh-treh-mohs

Break it here.
Unirlo aquí.
rrrohm-PEHR-loh ah-KEE

It must break on a joist.
El punto de unión de los extremos debe caer sobre una viga.
ehl POON-toh deh oo-NEE-ohn deh lohs
ex-treh-ohs deh-BEH OO-nah BEE-gah

Adding backing **It is too short, add backing.**
Es muy corto, añadir respaldo.
ehs MOO-ee KOHR-toh, ahn-yah-DEER rrreh-SPAHL-doh

Nail a two by four to the joist.
Clavar un dos por cuatro a la viga.
klah-BAHR oon dohs pohr KWAH-troh ah lah BEE-gah

It must be flush with the top.
Debe quedar a ras arriba.
DEH-beh keh-DAHR ah rrrahs-SAHN-teh ah-RRREE-bah

Bending joists

Bend the joist.
Doblar la viga.
doh-BLAHR lah BEE-gah

Push it with your foot.
Empujarla con el pie.
ehm-poo-HAHR-lah kohn ehl pyeh

Nail it.
Clavarla.
klah-BAHR-lah

Gluing sheathing

Glue the joists.
Pegar las vigas.
pah-GAHR lahs BEE-gahs

(Start / stop) the glue here.
(Empezar a / terminar de) pegar aquí.
(ehm-peh-SAHR ah / tehr-mee-NAHR deh)
peh-GAHR ah-KEE

Don't get glue under the tongue.
No untar pegamento debajo de la lengüeta.
noh oon-TAHR PEH-gah-mehn-toh deh-BAH-hoh
deh lah lehn-GWEH-tah

Placing the sheathing

Flip it over.
Voltearlo.
bohl-teh-AHR-loh

Put it exactly on the line.
Ponerlo exactamente en la línea.
poh-NEHR-loh ehk-sahk-tah-MEHN-teh ehn lah LEEN-yah

Break it here.
Unirlo aquí.
rrrohm-PEHR-loh ah-KEE

Nail the corners.
Clavar las esquinas.
klah-BAHR lahs eh-SKEE-nahs

Installing tongue and groove

Slide it in.
Deslizarlo.
dehs-lee-SAHR-loh

Stand (here / there).
Levantar (aquí / allá).
leh-bahn-TAHR (ah-KEE / ah-YAH)

Using the sledge-hammer

Hit it with the sledge-hammer.
Golpearlo con el acotillo.
gohl-peh-AHR-loh kohn ehl ah-koh-TEE-yoh

Protect the tongue with a board.
Proteger la lengüeta con una tabla.
pro-teh-HEHR lah lehn-GWEH-tah kohn oon TAH-blah

Nailing corners

Nail the corners.
Clavar las esquinas.
klah-BAHR lahs eh-SKEE-nahs

Nail it three inches back from the edge.
Clavarlo a tres pulgadas del borde.
klah-BAHR-loh ah trehs pool-GAH-dahs dehl BOHR-deh

Cutting sheathing

Cut it on the stack.
Cortarlo en la pila.
kohr-TAHR-loh ehn lah PEE-lah

Don't cut the plywood below.
No cortar la madera terciada más abajo.
noh kohr-TAHR ehl plywood deh-BAH-hoh

Cut it to (forty eight) inches.
Cortarlo de (cuarenta y ocho) pulgadas.
kohr-TAHR-loh deh (kwah-REHN-tah ee OO-choh)
pool-GAH-dahs

Marking joists	**Mark the joists at the edge.** Marcar las vigas en el borde. *mahr-KAHR lahs BEE-gahs ehn ehl BOHR-deh*
Nailing sheathing	**Nail the sheathing.** Clavar el entablado. *klah-BAHR ehl ehn-tah-blah-doh* **Use (eights / ring shanks).** Usar (ochos / clavos con fuste corrugado). *oo-SAHR (OH-chohs / KLAH-bohs kohn* *FOO-steh koh-rrroo-GAH-doh)* **Sink the nails.** Hundir los clavos. *hoon-DEER lohs KLAH-bohs*
Flush-nailing	**Use the flush-nailer.** Usar la clavadora parejo. *oo-SAHR lah klah-bah-DOH-rah pah-REH-hoh* **Turn the pressure (down / up). More.** (Subir / bajar) la presión. Más. *(soo-BEER / bah-HAHR) lah preh-SYOHN. Mahs*
Snapping perimeter lines	**Snap a line at the edge.** Marcar una línea en el borde con tiza. *mark-KARH OO-nah LEEN-yah ehn* *ehl BOHR-deh kohn TEE-zah*
Cutting the perimeter	**Cut the outside edge.** Cortar el borde exterior. *kohr-TAHR ehl BOHR-deh ehk-stehr-YOHR* **Cut more here.** Cortar más aquí. *kohr-TAHR mahs ah-KEE*
Pulling shiners (missed nails)	**Pull the nails that missed.** Sacar los clavos que no dieron. *sah-KAHR lohs KLAH-bohs keh noh dyeh-ROHN*

SETTING WINDOWS

Set the windows.
Cuadrar las ventanas.
kwah-DRAHR lahs behn-TAH-nahs

Papering openings

Put paper around the opening.
Poner papel alrededor de la abertura.
*poh-NEHR pah-PEHL ahl-rrreh-deh-DOHR
deh lah ah-behr-TOO-rah*

Laying out windows

Lay out the windows.
Colocar las ventanas.
koh-loh-KAHR lahs behn-TAH-nahs

Put them at their proper locations.
Ponerlas en la ubicación correcta.
*poh-NEHR-lahs ehn lah
oo-bee-kah-SYOHN koh-RRREHK-tah*

Put them (outside / inside).
Ponerlas (afuera / adentro).
poh-NEHR-lahs (ah-FWEH-rah / ah-DEHN-troh)

Heavy windows

Get some help.
Conseguir ayuda.
kohn-seh-GEER ah-YOO-dah

Use (four) men.
Usar (cuatro) hombres.
oo-SAHR (KWAH-troh) OHM-brehs

Placing windows

Put it in the opening.
Ponerla en la abertura.
poh-NEHR-lah ehn lah ah-behr-TOO-rah

Put the bottom in first.
Poner primero la parte inferior.
poh-NEHR pree-MEH-roh lah PAHR-teh een-fehr-YOHR

The opening is too small.
La abertura es demasiado pequeña.
lah ah-behr-TOO-rah ehs deh-mah-SYAH-doh
peh-KEHN-yah

Centering the window

Center it in the opening.
Centrarla en la abertura.
sehn-TRAHR-lah ehn lah ah-behr-TOO-rah

…this way / a little more / too much
…esta dirección / un poco más / demasiado
…EH-stah dee-rehk-SYOHN / oon
POH-koh mahs / deh-mah-SYAH-doh

Leveling the window

Level the window.
Nivelar la ventana.
nee-beh-LAHR la behn-TAH-nah

Tacking the corners

Tack the corners.
Fijar las esquinas.
fee-HAHR lahs eh-SKEE-nahs

Operating the window

Operate the window.
Operar la ventana.
oh-pehr-RAHR lah behn-TAH-nah

(Lift / shim) it here.
(Levantarla / cuñarla) aquí.
(leh-bahn-TAHR-lah / koon-YAHR-lah) ah-KEE

Straightening long windows

Straighten the window.
Enderezar la ventana.
ehn-deh-reh-SAHR lah behn-TAH-nah

The (top / bottom) must be straight.
La parte (superior / inferior) debe estar alineada.
lah PAHR-teh (soo-pehr-YOHR / een-fehr-YOHR)
DEH-beh eh-STAHR ah-lee-neh-AH-dah

(The tops / the bottoms) must align.
Las partes (superiores / inferiores) deben alineadas.
lahs PAHR-tehs (soo-pehr-YOH-rehs /
een-fehr-YOH-rehs) DEH-behn ah-lee-neh-AH-dahs

Aligning multiple windows (vertically)	**It must align with the one (above / below).** Debe alineada con la que está (arriba / abajo). *DEH-beh ah-lee-neh-AH-dah kohn lah keh EH-stah* *(ah-RRREE-bah / ah-BAH-hoh)*
Spacing multiple windows	**The spaces must be equal.** Los espacios deben ser iguales. *lohs eh-SPAH-syohs DEH-behn sehr ee-GWAH-lehs*
Nailing flanges	**Nail the outside.** Clavar el exterior. *klah-BAHR ehl ehk-stehr-YOHR* **Nail every (six / twelve / sixteen) inches.** Clavar cada (seis / doce / dieciséis) pulgadas. *klah-BAHR KAH-dah (sais / DOH-seh /* *dyehs-EE-SAIS) pool-GAH-dahs*
Nails	**Use (roofing / galvanized) nails.** Usar clavos (para techo / galvanizados). *oo-SAHR KLAH-bohs (PAH-rah TEH-choh /* *gahl-bah-nee-SAH-dohs)* **...teco nails / eights** ...clavos tecos / ochos *...KLAH-bohs TEE-kohs / OH-chohs*
Nailing through the sides	**Nail through the sides.** Clavar a través de los lados. *klah-BAHR ah trah-BEHS deh lohs LAH-dohs* **Use finish nails.** Usar clavos de acabado. *oo-SAHR KLAH-bohs deh ah-koh-BAH-doh* **Use the gun.** Usar la clavadora. *oo-SAHR lah klah-bah-DOH-rah*

■ SETTING SECOND FLOOR WINDOWS

Laying out second floor window

Lay out the windows for the (second / third) floor.
Colocar las ventanas para el (segundo / tercer) piso.
koh-lo-KAHR lahs behn-TAH-nahs PAH-rah ehl
(seh-GOON-doh / tehr-SEHR) PEE-soh

Setting windows from pump jacks

Put it on the pump jack.
Ponerla en el caballete de bombeo.
poh-NEHR-lah ehn ehl kah-bah-YEH-teh deh bohm-BEE-yoh

Setting windows from scaffolding

Take it out on the scaffolding.
Sacarla en el andamiaje.
sah-KAHR-lah ehn ehl ahn-dah-MYAH-heh

Placing the window

Slide it out through the opening.
Deslizarla a través de la abertura.
dehs-lee-SAHR-lah ah trah-BEHS deh lah ah-behr-TOO-rah

Do you have it?
¿La tiene?
lah TYEHN-eh

■ SETTING DOORS (EXTERIOR)

Caulking the threshold

Caulk the threshold.
Calafatear el umbral.
kah-lah-fah-teh-AHR ehl oom-BRAHL

Plumbing the hinge side

Remove the (pins / door).
Remover (las clavijas / la puerta).
rrreh-moh-BEHR (lahs klah-BEE-has / lah PWEHR-tah)

Plumb this side first.
Enderezar este lado primero.
ehn-deh-reh-SAHR EH-steh LAH-doh pree-MEH-roh

Fastening the jamb

Nail it here.
Clavarla aquí.
klah-BAHR-lah ah-KEE

Nail it behind the weatherstrip.
Clavarla detrás del burlete.
klah-BAHR-lah deh-TRAHS dehl boor-LEH-teh

Replace this screw with a longer one.
Reemplazar este tornillo con uno más largo.
reh-ehm-plah-SAHR EH-steh tohr-NEE-yoh
kohn OO-noh mahs LAHR-goh

Use the nail gun.
Usar la clavadora automática.
oo-SAHR lah klah-bah-DOH-rah ah-toh-MAH-tee-kah

SETTING DOORS (INTERIOR)

Plumbing the hinge side

Remove (the pins / the door).
Remover (las clavijas / la puerta).
rrreh-moh-BEHR (lahs klah-BEE-has / lah PWEHR-tah)

Plumb this side first.
Enderezar este lado primero.
ehn-deh-reh-SAHR EH-steh LAH-doh pree-MEH-roh

Re-hanging the door

Put the door back in.
Volver a poner la puerta.
bohl-BEHR ah poh-NEHR lah PWEHR-tah

Put the pins in.
Poner las clavijas.
poh-NEHR lahs klah-BEE-has

Setting the other side

Look at the crack.
Mirar la grieta.
mee-RAHR lah gree-EH-tah

It must be the same at the top and sides.
Debe ser igual arriba y en los lados.
DEH-beh sehr ee-GWAHL ah-RRREE-bah
ee ehn lohs LAH-dohs

Shimming the head	**(Nail / shim) the head.**
	(Clavar / cuñar) el cabezal.
	(klah-BAHR / koon-YAHR) ehl kah-bah-SEH-rah

Nailing	**Nail it here.**
	Clavarla aquí.
	klah-BAHR-lah ah-KEE

Use the finish gun.
Utilizar la máquina de acabado.
oo-tee-lee-SAHR lah mah-KEE-nah deh ah-kah-BAH-doh

CLEAN-UP

Removing braces	**Remove the braces.**
	Sacar las riostras.
	ah-KAHR lahs rrree-OH-strahs

Scrapping out	**(Clean / sweep) the floor.**
	(Limpiar / barrer) el piso.
	(leem-PYAHR / bah-RRREHR) ehl PEE-soh

Keep the straight lumber.
Guardar la madera derecha.
gwahr-DAHR lah mah-DEH-rah deh-REH-chah

Keep wood longer than (twenty-four) inches.
Guardar la madera de más de (veinticuatro) pulgadas.
gwahr-DAHR lah mah-DEH-rah deh mahs deh
(behn-tee-KWAH-troh) pool-GAH-dahs

Cleaning lumber	**Clean the lumber.**
	Limpiar la madera.
	leem-PYAHR lah mah-DEH-rah

(Pull / bend over) the nails.
(Sacar / doblar) los clavos.
sah-KAHR / doh-BLAHR) lohs KLAH-bohs

Stacking lumber	**Stack the good lumber in front.**
	Apilar la madera buena enfrente.
	ah-pee-LAHR lah mah-DEH-rah BWEH-nah ehn-FREHN-teh

Put it on stickers.
Ponerla en los soportes.
poh-NEHR-lah ehn lahs soh-POHR-tehs

orting lumber

Separate the lumber by size.
Separar la madera por tamaños.
seh-pahr-RAHR lah mah-DEH-rah pohr tah-MAHN-yohs

crap pile location

Put the scrap pile (here / there / in front).
Poner la pila de desperdicios (aquí / allá / al frente).
poh-NEHR lah PEE-lah deh dehs-pehr-DEE-syohs
(ah-KEE / ah-YAH / ahl FREHN-teh

Put the scrap in the box.
Poner los desperdicios en la caja.
poh-NEHR lohs dehs-pehr-DEE-syohs ehn lah KAH-hah

SCAFFOLDING

Scaffolding location

Build a scaffold down the middle of the floor.
Construir un andamiaje en el medio del piso.
kohn-stroo-EER oon ahn-dah-MYAH-heh
ehn ehl MEH-dyoh deh PEE-soh

Scaffold height

Make it (two) inches lower than the walls.
Debe ser (dos) pulgadas más bajo que las paredes.
DEH-beh sehr (dohs) pool-GAH-dahs
mahs BAH-hoh keh lahs peh-REH-dehs

...(twelve) feet / inches off the floor
...(doce) pies / pulgadas sobre el piso
...(DOH-seh) pyehs / pool-GAH-dahs SOH-breh ehl PEE-soh

Plank material

Use (two by twelve) for planks.
Usar (dos por doce) para los tablones.
oo-SAHR (dohs pohr DOH-seh)
PAH-rah lohs tah-BLOH-nehs

BLOCKS

Cutting blocks

Cut (eighty) blocks.
Cortar (ochenta) bloques.
kohr-TAHR (oh-CHEHN-tah) BLOH-kehs

Cut them out of two by (four / six / ten).
Cortarlos de un dos por (cuatro / seis / diez).
kohr-TAHR-lohs deh oon dohs pohr
(KWAH-troh / sais / dyehs)

Pulling truss layout

Mark truss layout on the walls.
Haga el trazo de la armadura en las paredes.
ah-GAH ehl trah-zoh deh lah ahr-mah-DOO-rah

Follow the layout on the floor.
Seguir el trazado en el piso.
seg-GEER trah-ZAH-doh ehn ehl PEE-soh

They go on (twenty-four)-inch centers.
Se espacian(veinticuatro) pulgadas enter centros.
seh esh-pah-CYON (behn-tee-KWAH-troh)
pool-GAH-dahs ehn-TREH sehn-TROS

ROLLING TRUSSES

Strongback

Nail a two by (four / six) to the outside wall.
Clavar un dos por (cuatro / seis) a la pared exterior.
klah-BAHR oon dohs pohr (KWAH-troh / sais)
ah lah pah-REHD ehk-stehr-YOHR

Use a (sixteen) footer.
Usar un madero de (dieciséis) pies.
oo-SAHR oon mah-DEH-roh deh (dyehs-EE-sais) pyehs

Extend it (six) feet above the wall.
Extenderlo (seis) pies sobre la pared.
ehk-stehn-DEHR-loh (sais) pyehs SOH-breh lah pah-REHD

Laying out the first trusses

Nail the gabled end to the bracing.
Clavar la punta con aguilón al arriostramiento.
klah-BAHR lah POON-tah kohn ah-gwee-LOHN
ahl ah-rryohs-trah-MYEN-toh

Lean this one against the last one.
Recostar ésta contra la última.
rrreh-koh-STAHR eh-STAH KOHN-trah lah ool-TEE-mah

Lay it on top of the last one.
Recostarla sobre la última.
ah-koh-STAHR-lah SOH-breh lah ool-TEE-mah

Laying out trusses

Put them on layout.
Alinearlas con el trazo.
kohn ehl trah-ZOH

Laying out blocks

Lay out the blocks.
Colocar los bloques correctamente.
koh-loh-KARH lohs BLOH-kehs korrrh-ekth-ah-mehn-teh

Figure 7. Laying Out Trusses
           Colocación de las armaduras

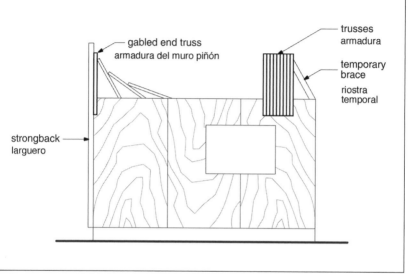

gabled end truss
armadura del muro piñón

trusses
armadura

temporary
brace

riostra
temporal

strongback
larguero

The gabled-end is tacked to the strongback and the trusses are placed on layout.

Put them in the middle.
Ponerlos en el medio.
poh-NEHR-lohs ehn ehl MEH-dyoh

Put a block on each truss.
Poner un bloque en cada armadura.
poh-NEHR oon BLOH-keh ehn KAH-dah ahr-mah-DOO-rah

Laying out trusses

Stand the trusses.
Parar las armaduras.
pah-RAHR lahs ahr-mah-DOO-rahs

Block (the top / the outside wall).
Bloquear (el tope / la pared exterior).
bloh-keh-AHR (ehl TOH-peh /
lah pah-REHD ehk-stehr-YOHR)

Align the trusses.
Alinear las armaduras.
ah-leen-YAHR lahs ahr-mah-DOO-rahs

Aligning trusses

It must be flush at the outside.
Debe quedar a ras con el exterior.
DEH-beh keh-DAHR ah rrrras-SAHN-teh
kohn ehl ehk-stehr-YOHR

Is it flush?
¿Está a ras?
EH-stah ah rrrras-SAHN-yeh

The other side too?
¿También el otro lado?
tahm-BYEHN ehl OH-troh LAH-doh

It needs to go (to you / to me / in / out).
Necesita ir (hacia usted / hacia mí / adentro / afuera).
neh-seh-SEE-tah eer (AH-syah oo-STEHD /
AH-syah MEE / ah-DEHN-troh / ah-FWEH-rah)

Sway (diagonal) brace

Run a diagonal brace back to the wall.
Instalar una riostra diagonal hasta la pared.
een-stah-LAHR OO-nah rrree-OH-strah
dyah-goh-NAHL ah-stah lah pah-REHD

Nail the bottom of the brace to the wall.
Clavar la parte inferior de la riostra a la pared.
klah-BAHR lah PAHR-teh een-fehr-YOHR
deh lah rrree-OH-strah ah lah pah-REHD

Checking layout

Check the layout.
Revisar las medidas y trazos.
reh-bee-SAHR lahs meh-dee-DAH ee trah-ZOHS

They must be within a (half) inch.
No deben exceder (media) pulgada.
DEH-behn eh-STAHR ah MEH-nohs de
(MEH-dyah) pool-GAH-dah

Cut a special block.
Cortar un bloque especial.
kohr-TAHR oon BLOH-keh eh-speh-SYAHL

HIPS AND VALLEYS

Laying out the hips / valleys

Lay out the (hips / valleys).
Poner las limas (tesas / hoyas).
poh-NEHR lahs LEE-mahs (TEH-sahs / OH-yahs)

Lean them up at the corners.
Recostarlas en las esquinas.
rrreh-koh-STAHR-lahs ehn lahs eh-SKEE-nahs

Installing hips / valleys

It must be flush at the corner.
Debe quedar a ras con la esquina.
DEH-beh keh-DAHR ah rrras-SAHN-teh
kohn lah eh-SKEE-nah

Straightening hips and valleys

Sight the (hip / valley).
Visar las limas (tesas / hoyas).
bee-SAHR lahs LEE-mahs (TEH-sahs / OH-yahs)

Nail a brace in the middle.
Clavar una riostra en el medio.
klah-BAHR oon rrree-OH-strah ehn ehl MEH-dyoh

It must go (in / out).
Debe ir (adentro / afuera).
DEH-beh eer (ah-DEHN-troh / ah-FWEH-rah)

…tack it / nail it
…fijarla / clavarla
…fee-HAHR-lah / klah-BAHR-lah

Laying out jacks

Lay out the jacks.
Colocar los cabrios cortos.
koh-loh-KARH lohs kah-BREE-ohs KOHR-tohs

Hanging jacks

Hang the jacks.
Instalar los cabrios cortos.
ihn-stah-larh lohs kah-BREE-ohs KOHR-tohs

Start a toenail in the top.
Fijar un clavo sesgado en el tope.
fee-HAHR oon KLAH-boh sehs-GAH-doh ehn ehl TOH-peh

Hang them on the wall.
Instalarlos en la pared.
ihn-stah-larh-lohs ehn lah pah-REHD

Nailing schedule

(Five) nails per jack.
(Cinco) clavos por cabrio.
(SEEN-koh) KLAH-bohs pohr kah-BREE-oh

Keep them straight.
Mantenerlos alineados.
mahn-teh-NEHR-lohs deh-REH-chohs

Checking for flush at the wall

They must be flush with the outside.
Deben quedar a ras con el exterior.
DEH-behn keh-DAHR ah rrras-SAHN-te kohn ehl ehk-stehr-YOHR

Blocking outside walls

Block the outside walls.
Bloquear las paredes exteriores.
bloh-keh-AHR lahs peh-REH-dehs ehk-stehr-YOH-rehs

Put it (outside / on top of) the wall.
Ponerlo (fuera / encima) de la pared.
poh-NEHR-loh (foo-erh-rah /
ehn-SEE-mah) deh lah pah-REHD

Make it flush on top.
Que quede a ras con el tope.
keh KEH-deh ah rrras-SAHN-teh kohn ehl TOH-peh

Installing vent blocks

Every (third) block is a vent block.
Cada (tercer) bloque es para ventilación.
KAH-dah (tehr-SEHR) BLOH-keh ehs
pah-rah behn-tee-lah-SYOHN

CATWALK AND BRACES

Marking layout out the catwalk / braces

Mark layout on the (catwalk / braces).
Marcarel trazado en (la pasarela / las riostras).
mahr-KAHR ehl trah-ZAH-oh ehn
(lah pah-sarh-ehl-ah / lahs rrree-OH-strahs)

Mark them (twenty-four) inches on center.
Marcarlos a (veinticuatro) pulgadas entre centros.
mahr-KAHR-lohs ah (behn-tee-KWAH-troh)
pool-GAH-dahs ehn-TREH sehn-TROS

Laying out catwalk

Lay out the (catwalk / braces).
Colocar (la pasarela / las riostras).
koh-loh-KARH (lah pah-sarh-ehl-ah / lahs rrree-OH-strahs)

Installing catwalk

Install the catwalk.
Instalar la pasarela.
een-stah-LAHR lah pah-sarh-ehl-ah

Put a row here (and here).
Poner una hilera aquí (y aquí).
poh-NEHR OO-nah ee-LEH-rah ah-KEE (ee ah-KEE)

Nailing catwalk

Nail the catwalk.
Clavar la pasarela.
klah-BAHR lah pah-sarh-ehl-ah

Keep it on layout.
Mantenerla de acuerdo con la disposición.
mahn-teh-NEHR-lah deh akh-oo-erh-doh
kohn lah dees-poh-see-SYOHN

Installing braces

Install the braces.
Instalar las riostras.
een-stah-LAHR lahs rrree-OH-strahs

Put a row at the (top / here).
Poner una hilera (en el tope / aquí).
poh-NEHR OO-nah ee-LEH-rah (ehn ehl TOH-peh / ah-KEE)

▐ BACKING

Backing location

Put backing here.
Poner el respaldo aqui.
poh-NEHR ehl rrreh-SPAHL-doh ah-KEE

Laying out backing

Lay out the backing.
Poner el respaldo.
poh-NEHR ehl rrreh-SPAHL-doh

Backing material

Use (two by four).
Usar (dos por cuatro).
oo-SAHR (dohs pohr KWAH-troh)

Watch, do it like this.
Mire, así.
MEE-reh ah-SEE

Cutting backing

Cut in the backing.
Cortar el respaldo.
kohr-TAR ehl rrreh-SPAHL-doh

Hanging backing

Hang the backing from the truss.
Colgar el respaldo desde la armadura.
kohl-GAHR ehl rrreh-SPAHL-doh
dehs-deh lah ahr-mah-DOO-rah

Nailing backing	**Nail the backing.** Clavar el respaldo. *klah-BAHR ehl rrreh-SPAHL-doh*
	Nail (the backing / the wall) to the blocks. Clavar (el respaldo / la pared) a los bloques. *klah-BAHR (ehl rrreh-SPAHL-doh / lah pah-REHD)* *ah lohs BLOH-kehs*
	Watch, do it like this. Mire, así. *MEE-reh ah-SEE*
Backing block locations	**Block this bay.** Bloquear este vano. *bloh-keh-AHR EH-steh bah-noh*
	Nail into the wall below. Clavar en la pared abajo. *klah-BAHR ehn lah pah-REHD ah-BAH-hoh*
Truss clips	**Install the truss clips.** Instalar los clips de armaduras. *een-stah-LAHR lohs clips deh ahr-mah-DOO-rahs*
	Put one on each (truss / wall). Poner uno en cada (armadura / pared). *poh-NEHR OO-noh ehn KAH-dah* *(ahr-mah-DOO-rah / pah-REHD)*
Hurricane clips	**Install the hurricane clips.** Instalar los clips de huracán. *een-stah-LAHR los clips deh oo-rah-KAHN*
	Watch, do it like this. Mire, así. *MEE-reh ah-SEE*

Figure 8. Roof Framing Members
Elementos de la estructura del techo

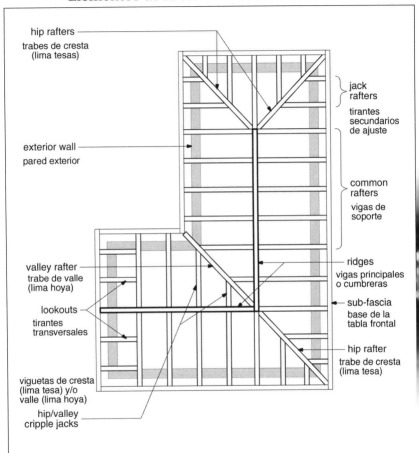

hip rafters
trabes de cresta
(lima tesas)

jack
rafters
tirantes
secundarios
de ajuste

exterior wall
pared exterior

common
rafters
vigas de
soporte

valley rafter
trabe de valle
(lima hoya)

ridges
vigas principales
o cumbreras

lookouts
tirantes
transversales

sub-fascia
base de la
tabla frontal

hip rafter
trabe de cresta
(lima tesa)

viguetas de cresta
(lima tesa) y/o
valle (lima hoya)

hip/valley
cripple jacks

Framing members of a typical conventionally framed roof.

CUTTING RAFTERS (PRODUCTION METHOD)

Setting up horses

Put the horses here.
Poner las caballos aquí.
poh-NEHR lohs kah-BAH-yohs ah-KEE

Put them (twelve) feet apart.
Ponerlos cada (doce) pies.
poh-NEHR-lahs KAH-dah (DOH-seh) pyehs

Level the horses.
Nivelar las caballos.
nee-beh-LAHR lohs kah-BAH-yohs

Loading horses

Load the horses.
Cargar las caballos.
kahr-GAHR lohs kah-BAH-yohs

Put them on edge.
Ponerlos en el borde.
poh-NEHR-lohs ehn ehl BOHR-deh

Rafter material size

Load two by (six / eight / ten / twelve).
Cargar dos por (seis / ocho / diez / doce).
kahr-GAHR dohs pohr (sais / OH-choh / dyehs / DOH-seh)

Crowning rafters

Crown the rafters.
Coronar los cabrios.
koh-roh-NAHR lohs kah-BREE-ohs

Put the crowns down.
Poner las coronas hacia abajo.
poh-NEHR lahs koh-ROH-nahs ah-BAH-hoh

Rafter layout (top cut)

Mark the top cut.
Marcar el corte superior.
mahr-KAHR-ehl KOHR-teh soo-pehr-YOHR

Make it (twenty-two and one-half).
Hacerlo de (veintidós y medio).
ah-SEHR-loh deh (behn-tee-DOHS ee MEH-dyoh)

Make it (five) and twelve.
Hacerlo de (cinco) y doce.
ah-SEHR-loh deh (SEEN-koh) ee DOH-seh

Rafter layout (seat cut)

Mark the seat cut.
Marcar el corte inferior.
mahr-KAHR ehl KOHR-teh een-fehr-YOHR

Watch, do it like this.
Mire, así.
MEE-reh, ah-SEE

Snapping lines

Snap cut-lines across the rafters.
Marcar con tiza las líneas de corte sobre los cabrios.
mark-KAHR kohn TEE-sah lahs LEEN-yahs deh
KOHR-teh SOH-breh lohs KAH-bree-yohs

Snap a line here.
Marcar una línea con tiza aquí.
mark-KAHR OO-nah LEEN-yah kohn TEE-sah ah-KEE

Cutting seat cuts

Cut the seat cut first.
Cortar el corte inferior primero.
kohr-TAHR ehl KOHR-teh een-fehr-YOHR pree-MEH-roh

Cut along this line.
Cortar por esta línea.
kohr-TAR pohr EH-stah LEEN-yah

Set the saw to (thirty-two / fifty-eight).
Ajustar la sierra a (treinta y dos / cincuenta y ocho).
ah-hoo-STAHR lah SYEH-rrrah ah (TRAIN-tah ee dohs /
seen-KWEHN-tah ee OH-choh) GRAH-dohs

Set the blade depth.
Ajustar la profundidad de la cuchilla.
ah-hoo-STAHR lah proh-foon-dee-DAHD
deh lah koo-CHEE-ah

Cutting top cuts

Cut the top cut.
Cortar el corte superior.
kohr-TAHR ehl KOHR-teh soo-pehr-YOHR

Check the angle.
Revisar el ángulo.
rrreh-bee-SAHR ehl ahn-GOO-loh

Watch, do it like this.
Mire, así.
MEE-reh ah-SEE

Cutting rafter blocks | **Cut (seventy) blocks.**
Cortar (setenta) bloques.
kohr-TAHR (sah-TEHN-tah) BLOH-kehs

Cut them (twenty-two and three-eighths) inch(es).
Cortarlos (veintidós y tres octavos de) pulgada(s).
kohr-TAHR-lohs (behn-tee-DOHS ee trehs
ohk-TAH-bohs deh) pool-GAH-dahs

Block material size | **Use two by (four / six / eight / ten) for blocks.**
Usar dos por (cuatro / seis / ocho / diez) para los bloques.
oo-SAHR dohs pohr (KWAH-troh / sais / OH-choh /
dyehs) PAH-rah lohs BLOH-kehs

CUTTING RAFTERS USING A JIG

Cut the rafters.
Cortar los cabrios.
kohr-TAHR lohs kah-BREE-ohs

Use the jig.
Usar la sierra de vaivén.
oo-SAHR ehl jig

Loading horses | **Stack the two by (eight / ten / twelve) in stacks of (twelve).**
Apilar los dos por (ocho / diez / doce) en pilas de (doce).
ah-pee-LAHR lohs dohs pohr (OH-choh / dyehs /
DOH-seh) ehn PEE-lahs deh (DOH-seh)

Lay them flat.
Recostarlos.
ah-koh-STAHR-lohs

| Crowning rafters | **Crown them.**
| | Coronarlos.
| | *koh-roh-NAHR-lohs*

Put all the crowns on the same side.
Poner todas las coronas en el mismo lado.
ppoh-NEHR TOH-dohs lahs koh-ROH-nahs
ehn ehl MEES-moh LAH-doh

Top cut

Mark and cut the top cut first.
Marcar y cortar el corte superior primero.
mahr-KAHR ee kohr-TAHR ehl
KOHR-teh soo-pehr-YOHR pree-MEH-roh

Cut off the split end.
Cortar la punta larga.
kohr-TAHR lah POON-tah LArh-gah

**Pulling
measurements**

Hook the long point.
Enganchar el punto largo.
ehn-gahn-CHAHR ehl POON-toh LAHR-goh

Mark it at (one hundred forty-two and three-eighths).
Marcarlo a (ciento cuarenta y dos y tres octavos).
mahr-KAHR-loh ah (SYEHN-toh kwah-REHN-tah
ee dohs ee trehs ohk-TAH-bohs

**Using the
marking jig**

Watch, do it like this.
Mire, así.
MEE-reh ah-SEE

Put the jig on the mark.
Poner la sierra vaivén en la marca.
poh-NEHR ehl jig ehn lah MAHR-kah

Scribe the jig.
Calcar con la sierra vaivén.
kahl-KAHR kohn ehl jig

Marking tails

Mark the tail.
Marcar la cola.
mahr-KAHR lah KOH-lah

Make it (thirty six) inches long.
Que tenga (treinta y seis) pulgadas de largo.
keh TEHN-gah (TRAIN-tah ee sais)
pool-GAH-dahs deh LAHR-goh

Cutting rafters

Cut the bottom cut and tail.
Cortar el corte inferior y la cola.

Cut past the mark (one) inch.
Cortar (una) pulgada más allá de la marca.
kohr-TAHR (OO-nah) pool-GAH-dah mahs
ah-YAH deh lah MAHR-kah

FRAMING THE ROOF
SCAFFOLDING

Scaffolding location

Build a scaffold down the middle of the floor.
Construir un andamiaje en el medio del piso.
kohn-stroo-EER oon ahn-dah-MYAH-heh
ehn ehl MEH-dyoh deh PEE-soh

Scaffold height

...(twelve) feet / inches off the floor
...(doce) pies / pulgadas sobre el piso
...(DOH-seh) pyehs / pool-GAH-dahs SOH-breh ehl PEE-soh

Plank location

Use (one / two) rows of planks.
Usar (una / dos) hileras de tablones.
oo-SAHR (OO-nah / dohs) ee-LEH-rahs deh tah-BLOH-nehs

Put the planks (twelve) inches from the center.
Poner los tablones a (doce) pulgadas del centro.
poh-NEHR lohs tah-BLOH-nehs ah (DOH-seh)
pool-GAH-das dehl SEHN-troh

Plank material

Use (two by twelve) for planks.
Usar (dos por doce) para los tablones.
oo-SAHR (dohs pohr DOH-seh) PAH-rah lohs tah-BLOH-nehs

Laying out rafters

Lay out the rafters.
Colocar los cabrios.
koh-loh-KAHR lohs KAH-bree-yohs

Put the square end (up / down).
Poner la punta cuadrada hacia (arriba / abajo).
poh-NEHR lah POON-tah kwah-DRAH-dah AH-syah
(ah-RRREE-bah / ah-BAH-hoh)

Rafter locations

These rafters go (here/ there).
Estos cabrios van (aquí / allá).
EH-stohs KAH-bree-yohs bahn (ah-KEE / ah-YAH)

...on (the first / the second / the third) floor
...en el (primer / segundo / tercer) piso
...ehn ehl (pree-MEHR / seh-GOON-doh /
tehr-SEHR) PEE-soh

...in the middle / at the end / at the corner
...en el medio / en el extremo / en la esquina
...ehn ehl MEH-dyoh / ehn ehl ehk-STREH-moh /
ehn lah eh-SKEE-nah

...in front / in back
...adelante / atrás
...ah-deh-LAHN-teh / ah-TRAHS

Pulling up rafters

Pull up the rafters.
Sacar los cabrios.
sah-KAHR lohs KAH-bree-yohs

Put them (here / there).
Ponerlos (aquí / allá).
poh-NEHR-lohs (ah-KEE / ah-YAH)

Laying out ridge

Lay out the ridge.
Colocar el caballete.
koh-loh-KAHR ehl kah-bah-YEH-teh

Ridge location

That ridge goes (here/ there).
Ese caballete va (aquí / allá).
EH-seh kah-bah-YEH-teh bah (ah-KEE / ah-YAH)

...on the (first / second / third) floor
...en el (primer / segundo / tercer) piso
...en el (pree-MEHR / seh-GOON-doh / tehr-SEHR) piso

...in the middle / at the end
...en el medio / en la punta
...ehn ehl MEH-dyoh / ehn lah POON-tah

...in front / back
...adelante / atrás
...ah-deh-LAHN-teh / ah-TRAHS

Pulling up the ridge

Pull up the ridge.
Sacar el caballete.
sah-KAHR ehl kah-beh-YEH-teh

Put it in the middle.
Ponerlo en el medio.
poh-NEHR-loh ehn ehl MEH-dyoh

It must extend (four) feet past the wall.
Debe extenderse más allá de (cuatro) pies que la pared.
DEH-beh ehk-stehn-DEHR-seh mahs ah-YAH deh
(KWAH-troh) pyehs keh lah pah-REHD

Crowning the ridge

Crown the ridge.
Coronar el caballete.
koh-roh-NAHR ehl kah-bah-YEH-teh

The crown must be up.
La corona debe quedar hacia arriba.
lah koh-ROH-nah DEH-beh keh-DAHR
AH-syah ah-RRREE-bah

Pulling rafter layout

Mark layout on the ridge.
Marcar los trazos en el caballete.
mahr-KAHR lohs TRAH-sohs ehn ehl kah-bah-YEH-teh

Mark them (twenty-four) inches on center.
Marcarlos a (veinticuatro) pulgadas centradas.
mahr-KAHR-lohs ah (behn-tee-KWAH-troh)
pool-GAH-dahs sehn-TRAH-dahs

Layout is marked on the floor.
El trazo está en el piso.
Ehl TRAH-soh eh-STAH ehn ehl PEE-soh

Laying out braces	**Lay out (two by four) for braces.** Colocar (dos por cuatro) para riostras. *koh-loh-KAHR (dohs pohr KWAH-troh)* *PAH-rah rrree-OH-strahs*
	Put a brace at each end. Poner una riostra en cada punta. *poh-NEHR OO-nah rrree-OH-strah ehn KAH-dah POON-tah*
Nailing end commons	**Raise the last two rafters.** Levantar los dos últimos cabrios. *leh-bahn-TAHR lohs dohs ool-TEE-mohs KAH-bree-yohs*
	Hold them together at the top. Sostenerlos juntos en el tope. *soh-steh-NEHR-lohs HOON-tohs ehn ehl TOH-peh*
	Nail them at the wall. Clavarlos en la pared. *klah-BAHR-lohs ehn lah pah-REHD*
Setting the ridge	**Set the ridge.** Instalar el caballete. *een-stah-LAHR ehl kah-bah-YEH-teh*
	Slide the ridge up between the rafters. Deslizar el caballete entre los cabrios. *dehs-lee-SAHR ehl kah-bah-YEH-teh* *EHN-treh lohs KAH-bree-yohs*
	Tack it / nail it. Fijarlo / clavarlo. *fee-HAHR-loh / klah-BAHR-loh*
Leveling the ridge	**Level the ridge.** Nivelar el caballete. *nee-beh-LAHR ehl kah-bah-YEH-teh*
	This end is (high / low). Esta punta está (alto / bajo). *EH-stah POON-tah EH-stah (AHL-toh / BAH-hoh)*

Bracing the ridge	**Brace the ridge.** Arriostrar el caballete. *ah-rree-oh-STRAHR ehl kah-bah-YEH-teh*
	Use a diagonal brace. Usar una riostra diagonal. *oo-SAHR OO-nah rrree-OH-strah dyah-goh-NAHL*
Ridge post	**Put a post under it.** Poner un poste debajo. *poh-NEHR oon POH-steh deh-BAH-hoh*
	It must be plumb. Debe quedar a plomo. *DEH-beh keh-DAHR ah PLOH-moh*
Nailing common rafters	**Nail the common rafters.** Clavar los cabrios comunes. *klah-BAHR lohs KAH-bree-yohs koh-MOO-nehs*
	It must be flush on top. Debe quedar a ras en el tope. *DEH-beh keh-DAHR ah rrrahs ehn ehl TOH-peh*
Nailing schedule	**(Five) nails per rafter.** (Cinco) clavos por cabrio. *(SEEN-koh) KLAH-bohs pohr KAH-bree-yoh*
Straightening the ridge	**Straighten the ridge.** Enderezar el caballete. *ehn-deh-reh-SAHR ehl kah-bah-YEH-teh*
	Hold the rafter at the bottom. Sostener el cabrio abajo. *soh-steh-NEHR ehl KAH-bree-yoh ah-BAH-hoh*
	Move it (in / out). Moverlo hacia (adentro / afuera). *moh-BEHR-loh AH-syah (ah-DEHN-troh, ah-FWEH-rah)*
	(Tack / nail) it. Fijarlo / clavarlo. *fee-HAHR-loh / klah-BAHR-loh*

Marking the ridge for cutting	**Mark the ridge for the cut.** Marcar el caballete para el corte. *mahr-KAHR ehl kah-bah-YEH-teh PAH-rah ehl KOHR-teh*
	(Butt / hook) the wall. (Tocar / enganchar) la pared. *(toh-KAHR / ehn-gahn-CHAHR) lah pah-REHD*
	Mark (one hundred thirty) inches from the wall. Marcar a (ciento treinta) pulgadas de la pared. *mahr-KAHR ah (SYEHN-toh TRAIN-tah)* *pool-GAH-dahs deh pah-REHD*
	Plumb up with the level. Enderezar con el nivel. *ehn-deh-reh-SAHR kohn ehl nee-BEHL*
Cutting the ridge	**Cut the ridge.** Cortar el caballete. *kohr-TAHR ehl kah-bah-YEH-teh*
Crowning hips and valleys	**Crown the (hip / valley).** Coronar las limas (tesas / hoyas). *koh-roh-NAHR lahs LEE-mahs (TEH-sahs / OH-yahs)*
	The crown must be up. La corona debe quedar hacia arriba. *lah koh-ROH-nah DEH-beh* *keh-DAHR AH-syah ah-RRREE-bah*
Laying out hips / valleys	**Lay out the (hips / valleys).** Colocar las limas (tesas / hoyas). *koh-loh-KOHR lahs LEE-mahs (TEH-sahs / OH-yahs)*
Hip / valley location	**That (hip / valley) goes there.** Esa lima (tesa / hoya) va aquí. *EH-sah LEE-mah (TEH-sahs / OH-yahs) bah ah-KEE*
	…in the middle / at the end …en el medio / en la punta *…ehn ehl MEH-dyoh / ehn lah POON-tah*

…in front / back
…adelante / atrás
…ah-deh-LAHN-teh / ah-TRAHS

Setting hips and valleys

Set the (hip / valley).
Instalar las limas (tesas / hoyas).
een-stah-LAHR lahs LEE-mahs (TEH-sahs / OH-yahs)

Make it flush at (the wall).
Que quede a ras en (la pared).
keh KEH-deh ah rrrahs ehn (lah pah-REHD)

Blending hips and valleys

The (hip / valley) must blend in.
Las limas (tesas / hoyas) deben armonizar.
lahs LEE-mahs (TEH-sahs / OH-yahs)
DEH-behn ahr-mohn-ee-SAHR

Sight along the tops of the rafters.
Visar los topes de los cabrios.
bee-SAHR lohs TOH-pehs deh lohs KAH-bree-yohs

Straightening hips and valleys

Straighten the (hip / valley).
Enderezar las limas (tesas / hoyas).
ehn-deh-reh-SAHR lahs LEE-mahs (TEH-sahs / OH-yahs)

Nail a brace in the middle.
Clavar una riostra en el medio.
klah-BAHR OO-nah rrree-OH-strah ehn ehl MEH-dyoh

Laying out the hip / valley

Mark layout on the (hip / valley).
Marcar el trazo en las limas (tesas / hoyas).
mahr-KAHR ehl TRAH-soh ehn
las LEE-mahs (TEH-sahs / OH-yahs)

Mark them (twenty-four) inches on center.
Marcarlas a (veinticuatro) pulgadas entre centros.
mahr-KAHR-lahs ah (behn-tee-KWAH-troh)
pool-GAH-dahs EHN-treh SEHN-trohs

Laying out jacks

Lay out the jacks.
Colocar los cabríos cortos.
koh-loh-KAHR lohs kah-BREE-ohs KOHR-tohs

Hanging jacks	**Hang the jacks on the wall.**
	Colgar los cabrios cortos en la pared.
	kohl-GAHR lohs KAH-bree-yohs
	KOHR-tohs ehn lah pah-REHD
Nailing jacks	**Nail the jacks.**
	Clavar los cabrios cortos.
	klah-BAHR lohs KAH-bree-yohs KOHR-tohs
	The seat cuts must be tight.
	Los cortes inferiores deben ser apretados.
	lohs KOHR-tehs een-fehr-YOHR-ehs
	DEH–behn sehr ah-preh-TAH-dohs
Laying out blocks	**Laying out the blocks.**
	Colocar los bloques.
	koh-loh-KAHR lohs BLOH-kehs
Hanging blocks	**Hang the blocks from the plates.**
	Colgar los bloques de las placas.
	kohl-GAHR lohs BLOH-kehs deh lahs PLAH-kahs
Blocking rafters	**Block the rafters.**
	Bloquear los cabrios.
	bloh-keh-AHR lohs KAH-bree-yohs
Block location	**The blocks go (outside / on) the wall.**
	Los bloques van (afuera de / en) la pared.
	lohs BLOH-kehs bahn (ah-FWEH-rah deh / ehn)
	lah pah-REHD
	Put the blocks in the middle of the wall.
	Poner los bloques en el medio de la pared.
	poh-NEHR lohs BLOH-kehs ehn ehl
	MEH-dyoh deh lah pah-REHD
Nailing schedules	**(Four) nails in the block.**
	(Cuatro) clavos en el bloque.
	(KWAH-troh) KLAH-bohs ehn ehl BLOH-keh
Backing location	**Put backing here.**
	Poner el respaldo aqui.
	poh-NEHR ehl rrreh-SPAHL-doh ah-KEE

Laying out backing	**Lay out the backing.** Colocar el respaldo. *koh-loh-KAHR ehl rrreh-SPAHL-doh* **Use two by (four / six) for backing.** Usar dos por (cuatro / seis) para el respaldo. *oo-SAHR dohs pohr (KWAH-troh / sais)* *PAH-rah ehl rrreh-SPAHL-doh*
Cutting backing	**Cut in the backing.** Cortar el respaldo. *kohr-TAHR ehl rrreh-SPAHL-doh*
Hanging backing	**Hang the backing from the wall.** Colgar el respaldo de la pared. *kohl-GAHR ehl rrreh-SPAHL-doh deh lah pah-REHD*
Nailing backing	**Nail the backing.** Clavar el respaldo. *klah-BAHR ehl rrreh-SPAHL-doh* **Nail it to the top of the wall.** Clavarlo al tope de la pared. *klah-BAHR-loh ahl TOH-peh deh lah pah-REHD*
Rafter tie location	**Put rafter ties (here / in the whole roof).** Poner las ataduras de cabrios (aquí / en todo el techo). *poh-NEHR lahs ah-tah-DOO-rahs deh KAH-bree-yohs* *(ah-KEE / ehn TOH-doh ehl TEH-choh)* **Put one every (two / four / six) feet.** Poner una cada (dos / cuatro / seis) pies. *poh-NEHR OO-nah KAH-dah* *(dohs / KWAH-troh / sais) pyehs*
Rafter tie material size	**Use two by (four / six / eight).** Usar dos por (cuatro / seis / ocho). *oo-SAHR dohs pohr (KWAH-troh / sais / OH-choh)* **Use (sixteen) footers.** Usar maderos de (dieciséis) pies. *oo-SAHR mah-DEH-rohs deh (dyehs-ee-SAIS) pyehs*

Installing rafter ties	**Install the rafter ties.**
	Instalar las ataduras de cabrios.
	een-stah-LAHR lahs ah-tah-DOO-rahs deh KAH-bree-yohs
Collar tie location	**Put collar ties (here / in the whole roof).**
	Poner las ataduras de collar (aquí / en todo el techo).
	poh-NEHR lahs ah-tah-DOO-rahs deh koh-LAHR
	(ah-KEE / ehn TOH-doh ehl TEH-choh)
	They go in the upper third of the roof.
	Van en el tercio superior del techo.
	bahn ehn ehl TEHR-syoh soo-pehr-YOHR dehl TEH-choh
Collar tie material	**Use (one by six / two by four).**
	Usar (uno por seis / dos por cuatro).
	oo-SAHR (OO-noh pohr sais / dohs pohr KWAH-troh)
Installing collar ties	**Install the collar ties.**
	Instalar las ataduras de collar.
	een-stah-LAHR lahs ah-tah-DOO-rahs deh koh-LAHR
	(Five) nails per side.
	(Cinco) clavos por lado.
	(SEEN-koh) KLAH-bohs por LAH-doh
Gable stud locations	**Put gable studs here.**
	Poner los montantes de aguilón aquí.
	poh-NEHR lohs mohn-TAHN-tehs
	deh ah-gwee-LOHN ah-KEE
Gable vent framing	**Frame a hole for the gable vent.**
	Construir un hueco para la ventilación de aguilón.
	kohn-stroo-EER oon WEH-koh PAH-rah lah
	behn-tee lah-SYOHN deh ah-gwee-LOHN
	Make it (eighteen) inches wide.
	Que tenga (dieciocho) pulgadas de ancho.
	keh TEHN-gah (dyehs-ee-OH-choh)
	pool-GAH-dahs deh AHN-choh
	…and (twenty-four) inches high
	…y (veinticuatro) pulgadas de alto
	…ee (behn-tee-KWAH-troh) pool-GAH-dahs deh AHL-toh

Install the hurricane clips.
Instalar los clips de huracán.
een-stah-LAHR lohs clips deh oo-ree-KAHN

Put one on each rafter.
Poner uno en cada cabrio.
poh-NEHR OO-noh ehn KAH-dah KAH-bree-yoh

Double clips on the (two / four) corner rafters.
Clips dobles en los (dos / cuatro) cabríos de esquina.
clips DOH-blehs ehn lohs (dohs / KWAH-troh)
kah-BREE-ohs deh eh-SKEE-nah

Watch, do it like this.
Mire, así.
MEE-reh, ah-SEE

PUMP JACKS

Put pump jacks here.
Poner los balancines aquí.
poh-NEHR lohs bah-lahn-SEE-nehs ah-KEE

Put the poles (three) (feet / inches) from the wall.
Poner los postes a (tres) (pies / pulgadas) de la pared.
poh-NEHR lohs POH-stehs ah (trehs)
(pyehs / pool-GAH-dah) deh lah pah-REHD

...(eighteen / twenty-two) feet apart
...(dieciocho / veintidós) pies aparte
...(dyehs-ee-OH-choh / behn-tee-DOHS) pyehs ah-PAHR-teh

Stand the pole.
Levantar el poste.
leh-bahn-TAHR ehl POH-steh

Put your foot against the bottom.
Poner su pie contra la parte inferior.
poh-NEHR soo pyeh KOHN-trah lah PAHR-teh een-fehr-YOHR

Nailing roof braces (doughnuts)	**Nail the brace.** Clavar la riostra. *klah-BAHR lah rrree-OH-strah*
	You must hit the rafter. Debe clavar en el cabrio. *DEH-beh klah-BAHR ehl KAH-bree-yoh*
Planking	**Put the planks on the pump jacks.** Poner los tablones en los balancines. *poh-NEHR lohs tah-BLOH-nehs ehn lohs bah-lahn-SEE-nehs*
	(Chain / nail) the overlap. (Encadenar / clavar) el traslapo. *(ehn-kah-deh-NAHR / klah-BAHR) ehl trahs-LAH-poh*
Lowering pump jacks	**Hold on.** Sostener. *soh-steh-NEHR*
	Watch, do it like this. Mire, así. *MEE-reh, ah-SEE*

CUTTING RAFTER TAILS

Marking overhang lengths	**Mark the overhang at (twenty-one and three-quarters).** Marcar el voladizo a (veintiuno y tres cuartos). *mahr-KAHR ehl boh-lah-DEE-soh ah* *(behn-tee-OO-noh ee trehs KWAR-tohs)*
	Your tape must be level. Su cinta debe estar nivelada. *soo SEEN-tah DEH-beh eh-STAHR nee-beh-LAH-dah*
Snapping lines	**Snap a line on the bottom of the rafters.** Marcar una línea con tiza en la parte inferior de los cabrios. *mahr-KAHR OO-nah LEEN-yah kohn TEE-sah ehn lah* *PAHR-teh een-fehr-YOHR deh lohs KAH-bree-yohs*

Marking tails	**Mark the tails for cutting.** Marcar las colas a cortar. *mahr-KAHR lahs KOH-lahs ah kohr-TAHR*
	Use the (jig / level / square). Usar (sierra de vaivén / nivel / escuadra). *oo-SAHR oon (lah SYEH-rrrah deh bahy-BEHN /* *nee-BEHL / ehs-kwah-DRAH)*
Cut angle	**Cut them (plumb / square).** Cortarlas (a plomo / cuadradas). *kohr-TAHR-lahs (ah PLOH-moh/ kwah-DRAH-dahs)*
	Mark it at (twenty-two and one-half). Marcarla a (veintidós y medio) grados. *mahr-KAHR-lah ah (behn-tee-DOHS* *ee MEH-dyoh) GRAH-dohs*
Cutting tails	**Cut the tails.** Cortar las colas. *kohr-TAHR lahs KOH-lahs*
	Cut them from (above / the ladder / the scaffold). Cortarlas desde (arriba / la escalera / el andamiaje). *kohr-TAHR-lahs DEHS-deh* *(ah-RRREE-bah / ehl ahn-dah-MYAH-heh)*
Sub-fascia	**Install the sub-fascia.** Instalar los sub-fascia. *een-stah-LAHR lohs sub-fascia*
Sub-fascia material size	**Use two by (six / eight / ten) for sub-fascia.** Usar dos por (seis / ocho / diez) para los sub-fascia. *oo-SAHR dohs pohr (sais / OH-choh / dyehs)* *PAH-rah lohs sub-fascia*
	Use the (fourteen / sixteen) footers. Usar los maderos de (catorce / dieciséis) pies. *oo-SAHR lohs mah-DEH-rohs deh* *(kah-TOHR-seh / dyehs-ee-SAIS) pyehs*

Laying out sub-fascia	**Lay out the sub-fascia.** Colocar los sub-fascia. *koh-loh-KAHR lohs sub-fascia*
	Put one every (twelve / eighteen) feet. Poner uno cada (doce / dieciocho) pies. *poh-NEHR OO-nah KAH-dah* *(DOH-seh / dyehs-ee-OH-choh) pyehs*
Sub-fascia break locations	**It must break on a rafter.** El punto de unión de los extremos debe caer sobre un cabrio. *Ehl POON-toh deh oo-nee-YOHN deh lohs* *ehk-STREH-mohs deh-BEH oon KAH-bree-yoh*

▬▬▬▬▬▬▬ FASCIA

Fascia material size	**The fascia is (one / two) by (eight / ten / twelve).** El fascia es de (uno / dos) por (ocho / diez / doce). *ehl fascia ehs deh (OO-noh / dohs) pohr* *(OH-choh / dyehs / DOH-seh)*
Laying out fascia	**Lay out the fascia.** Colocar los fascia. *koh-loh-KAHR lohs fascia*
	Put one every (sixteen) feet. Poner uno cada (dieciséis) pies. *poh-NEHR OO-noh KAH-dah (dyehs-ee-SAIS) pyehs*
Measuring fascia	**Measure the fascia.** Medir los fascia. *meh-DEER lohs fascia*
	Cut it it (one hundred forty) inches to the (short / long) point. Cortar ése (ciento cuarenta) pulgadas al punto (corto / largo). *kohr-TAHR EH-seh (SYEHN-toh kwah-REHN-tah)* *pool-GAH-dahs deh POON-toh (KOHR-toh / LAHR-goh)*

…(Short to short / short to long / long to long) point
…Punto (corto a corto / corto a largo / largo a largo)
…POON-toh (KOHR-toh ah KOHR-toh /
KOHR-toh ah LAHR-goh / LAHR-goh ah LAHR-goh)

ascia break
ocations

It must break on a rafter.
El punto de unión de los extremos debe caer sobre un cabrio.
Ehl POON-toh deh oo-nee-YOHN deh lohs
ehk-STREH-mohs deh-BEH oon KAH-bree-yoh

cribing fascia

Scribe the fascia.
Calcar los fascia.
kahl-KAHR lohs fascia

Hold it in place.
Sostenerlo en su lugar.
soh-steh-NEHR-loh ehn soo loo-GAHR

Is it good ?
¿Está bien?
EH-stah byehn

Mark the (long / short) point.
Marcar el punto (largo / corto).
mahr-KAHR ehl POON-toh (LAHR-goh / KOHR-toh)

utting fascia

Cut the fascia.
Cortar los fascia.
kohr-TAHR lohs fascia

That cut is (thirty / forty-five) degrees.
Ese corte es de (treinta / cuarenta y cinco) grados.
EH-seh KOHR-teh ehs deh (TRAIN-tah /
kwah-REHN-tah ee SEEN-koh) GRAH-dohs

e-cutting bad cuts

It is open on the (bottom / top).
Está abierto (abajo / arriba).
EH-stah ah-BYEHR-toh (ah-BAH-hoh / ah-RRREE-bah)

plicing fascia

Splice it (here / in the middle).
Empalmarlo (aquí / en el medio).
ehn-pahl-MAHR-loh lohs fascia (ah-KEE / ehn ehl MEH-dyoh)

Splice it in the middle of the (door / window).
Empalmarlo en el medio de la (puerta / ventana).
ehn-pahl-MAHR-loh ehn ehl MEH-dyoh deh lah
(PWEHR-tah / behn-TAH-nah)

Cut the splice at (thirty) degrees.
Cortar el empalme a (treinta) grados.
kohr-TAHR ehl ehm-PAHL-meh ah (TRAIN-tah) GRAH-dohs

Nailing fascia

Nail the fascia.
Clavar los fascia.
klah-BAHR lohs fascia

Nail it by hand.
Clavarlo a mano.
klah-BAHR-loh ah MAH-noh

Fascia nails

Use galvanized nails.
Usar clavos galvanizados.
oo-SAHR KLAH-bohs gal-bah-nee-SAH-dohs

Working with a partner

(Up / down) a little. More. Too much.
Un poco más (arriba / abajo). Más. Demasiado.
oon POH-koh mahs (ah-RRREE-bah / ah-BAH-hoh).
Mahs. Deh-mah-SYAH-doh

No hammer marks.
Sin marcas de martillo.
seen MAHR-kahs deh mahr-TEE-yoh

Gluing joints

Glue the joints.
Pegar las juntas.
peh-GAHR lahs HOON-tahs

Use more glue.
Usar más pegamento.
oo-SAHR mahs PEH-gah-MEHN-toh

Let it dry.
Dejarlo secar.
deh-HAHR-loh seh-KAHR

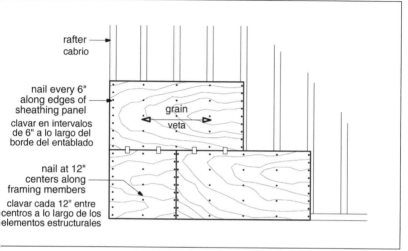

stallation details for roof sheathing.

Starter board	**Install the starter board.**
	Instalar la tabla inicial.
	een-stah-LAHR lah TAH-blah ee-nee-SYAHL

██████████ ROOF SHEATHING

Tying off	**You must be tied off.**
	Usted debe estar amarrado.
	oo-STEHD DEH-beh eh-STAHR ah-mah-RRRAH-doh
	You must wear a harness.
	Debe usar arreos.
	DEH-beh oo-SAHR ah-RRREH-ohs
Sheathing type	**Use (plywood / OSB) here.**
	Usar (madera terciada / OSB) aquí.
	oo-SAHR (mah-DEH-rah tehr-see-AH-dah / OSB) ah-KEE
	It must be stamped "exterior."
	Debe estar estampada "exterior."
	DEH-beh eh-STAHR eh-stahm-PAH-dah "ehk-stehr-YOHR"
Sheathing thickness	**Use the (half / five-eighths) inch plywood.**
	Usar madera terciada de (media / cinco octavos de) pulgada.
	oo-SAHR mah-DEH-rah tehr-see-AH-dah deh (MEH-dyah / SEEN-koh ohk-TAH-bohs deh) pool-GAH-dah
Preparing to load	**Nail one end to the top of the wall.**
	Clavar un extremo a la parte superior de la pared.
	klah-BAHR OO-nah oon ehk-STREH-moh ah lah pahr-teh deh lah pah-REHD
	Nail it to the side of the rafter.
	Clavarlo a un lado del cabrio.
	klah-BAHR-loh ah oon LAH-doh dehl KAH-bree-yoh
	Let them stick up (twenty) inches.
	Que sobresalgan (veinte) pulgadas.
	keh soh-breh-sahl-GAHN (BEHN-tee) pool-GAH-dahs

Load the roof with plywood.
Cargar el techo con madera terciada.
kahr-GAHR ehl TEH-choh kohn
mah-DEH-rah tehr-see-AH-dah

Get some help.
Conseguir ayuda.
kohn-seh-GEER ah-YOO-dah

Use (three) men.
Usar (tres) hombres.
oo-SAHR (trehs) OHM-brehs

Put it at (the top / the bottom) of the roof.
Ponerlo en la parte (superior / inferior) del techo.
poh-NEHR-loh ehn lah PAHR-teh
(soo-pehr-YOHR / een-fehr-YOHR) dehl TEH-choh)

Mark the end rafters.
Marcar los cabrios de extremo.
mahr-KAHR lohs KAH-bree-yohs deh ehk-STREH-moh

Hook the fascia.
Enganchar los fascia.
ehn-gahn-CHAHR lohs fascia

Butt the starter board.
Tocar la tabla inicial.
toh-KAHR lah TAH-blah ee-nee-SYAHL

Mark (forty-eight and one-half).
Marcar (cuarenta y ocho y medio).
mahr-KAHR (kwah-REHN-tah ee OH-choh ee MEH-dyoh)

Snap a line from corner to corner.
Marcar una línea con tiza de esquina a esquina.
mahr-KAHR OO-nah LEEN-yah kohn TEE-sah
deh eh-SKEE-nah ah eh-SKEE-nah

Re-snap this.
Marcar esto con tiza de nuevo.
mahr-KAHR EH-stoh kohn TEE-sah deh noo-EH-boh

Placing sheathing	**Put the plywood on the line.** Poner la madera terciada en la línea. *poh-NEHR lah mah-DEH-rah tehr-see-AH-dah* *ehn lah LEEN-yah*
	It must be on the line. Debe estar en la línea. *DEH-beh eh-STAHR ehn lah LEEN-yah*
	It must break on a rafter. El punto de unión de los extremos debe caer sobre un cabrio. *Ehl POON-toh deh oo-nee-YOHN deh lohs* *ehk-STREH-mohs deh-BEH oon KAH-bree-yoh*
	Break it here. Unirlo aquí. *oo-NEER-loh ah-KEE*
Tacking sheathing	**Nail the corners with eights.** Clavar las esquinas con ochos. *klah-BAHR lahs eh-SKEE-nahs kohn OH-chohs*
Spacing sheathing	**Leave a (one-eighth) inch gap between them.** Dejar un boquete de (un octavo) de pulgada entre ellas. *deh-HAHR oon boh-KEH-teh deh (oon ohk-TAH-boh)* *de pool-GAH-dah EHN-treh EH-yahs*
	...at the ends / all the way around ...en los extremos / todo alrededor *...ehn lohs ehk-STREH-mohs / TOH-toh ahl-rrreh-deh-DOHR*
	Tack an eight at the edge. Fijar un ocho en el borde. *fee-HAHR oon OH-choh ehn ehl BOHR-deh*
Marking rafters	**Mark the rafters at the edge.** Marcar los cabrios en el borde. *mahr-KAHR lohs KAH-bree-yohs ehn ehl BOHR-deh*
	Use keel (lumber crayon). Usar crayola de madera. *oo-SAHR kreh-OH-lah deh mah-DEH-rah*

aying out sheathing	**Lay out the next course.** Colocar la próxima hilada. *koh-loh-KARH lah prohk-SEE-mah ee-LAH-dah* **Lay out the whole row.** Colocar toda la hilera. *koh-loh-KARH TOH-doh lah ee-LEH-rah*
¹lacing breaks	**Come back to the next rafter.** Regresar al siguiente cabrio. *reh-greh-SAHR ahl see-GYEHN-teh KAH-bree-yoh* **Let them overlap.** Dejar que traslapen. *deh-HAHR deh trahs-lah-PEHN*
taggering breaks	**You must stagger the breaks.** Debe alternar los puntos de unión de los extremos. *DEH-beh ahl-tehr-NAR lohs POON-tohs deh* *oo-nee-YOHN deh lohs ehk-STREH-mohs* **Break it here.** Unirlo aquí. *oo-NEER-loh ah-KEE*
¹lywood clips	**Put a clip in each bay.** Poner un clip en cada vano. *poh-NEHR oon clip ehn KAH-dah BAH-noh* **Put it in the middle.** Ponerlo en el medio. *poh-NEHR-loh ehn ehl MEH-dyoh*
Cutting overlaps	**Snap a line down the center of the rafter.** Marcar una línea con tiza por el centro del cabrio. *mahr-KAHR OO-nah LEEN-yah kohn TEE-sah* *pohr ehl SEHN-troh dehl KAH-bree-yoh* **Cut through both of them.** Cortar a través de ambos. *kohr-TAHR ah trah-BEHS deh AHM-bohs*

Set the blade depth.
Ajustar la profundidad de la cuchilla.
ah-hoo-STAHR lah pro-foon
dee-DAHD deh lah koo-CHEE-yah

Cleaning the roof

Clean the roof.
Limpiar el techo.
leem-PYAHR ehl TEH-choh

Use the (blow-gun / broom).
Usar (el soplador / la escoba).
oo-SAHR (ehl soh-plah-DOHR / lah eh-SKOH-bah)

Use a piece of plywood.
Usar un pedazo de madera terciada.
oo-SAHR oon peh-DAH-soh deh mah-DEH-rah
tehr-see-AH-dah deh

Throwing scrap off the roof

Throw your scrap (here / there).
Tirar los desperdicios (aquí / allá).
tee-RAHR lohs des-per-DEE-syohs (ah-KEE / ah-YAH)

Before you throw, yell ("headache!").
Antes de tirarlos, gritar ("¡headache!").
AHN-tehs deh tee-RAHR-lohs, gree-TAHR (headache!)

Don't hit anybody.
No le pegue a nadie.
noh leh peh-GWEH ah NAH-dyeh

Nailing off the roof

Nail off the roof.
Clavar el techo.
klah-BAHR ehl TEH-choh

Use (eights / staples).
Usar (ochos / ganchos).
oo-SAHR (OH-chohs / GAHN-chohs)

Use the gun.
Usar la clavadora.
oo-SAHR lah klah-bah-DOH-rah

flush-nailer	**The nails must be flush.** Los clavos deben quedar parejos. *lohs KLAH-bohs DEH-beh keh-DAHR pah-REH-hohs*
	Use the flush nailer. Usar la clavadora parejo. *oo-SAHR lah klah-bah-DOH-rah pah-REH-hoh*
	Turn the pressure (up / down). (Subir / bajar) la presión. *(soo-BEER / BAH-hahr) lah preh-SYOHN*
Nailing schedules	**Nail the edges every (four / six) inches.** Clavar los bordes cada (cuatro / seis) pulgadas. *klah-BAHR lohs BOHR-dehs KAH-dah* *(KWAH-troh / sais) pool-GAH-dahs*
	Nail the middle every twelve inches. Clavar el medio cada doce pulgadas. *klah-BAHR ehl MEH-dyoh KAH-dah* *DOH-seh pool-GAH-dahs*
Checking nailing	**Check the nailing.** Revisar el clavado. *rrreh-bee-SAHR ehl klah-BAH-doh*
	You missed a row here. Le faltó una hilera aquí. *leh fahl-TOH OO-nah ee-LEH-rah ah-KEE*
	Sink the nails. Hundir los clavos. *hoon-DEER lohs KLAH-bohs*
Pulling shiners (missed nails)	**Pull the nails that missed.** Sacar los clavos que no dieron. *sah-KAHR lohs KLAH-bohs keh noh dyeh-ROHN*

CRICKETS

Build the cricket.
Construir el techo pequeño.
kohn-stroo-EER ehl TEH-choh poh-KEHN-yoh

SOFFITS (EXTERIOR)

Build the soffits.
Construir los sofitos.
kohn-stroo-EER lohs soh-FEE-tohs

Soffit locations

Put soffit (here / all the way around).
Poner el sofito (aquí / todo alrededor).
poh-NEHR ehl soh-FEE-toh (ah-KEE /
TOH-doh ahl-rrreh-deh-DOHR)

Soffit type

The soffit is (level / sloped).
El sofito está (nivelado / inclinado).
ehl soh-FEE-toh eh-STAH (nee-beh-LAH-doh /
een-klee-NAH-doh)

Nail the soffit to the bottom of the rafter tails.
Clavar el sofito a la parte inferior de las colas de los cabrios.
klah-BAHR ehl soh-FEE-toh ah lah PAHR-teh
een-fehr-YOHR deh lahs KOH-lahs deh lohs KAH-bree-yohs

Soffit layout

Level across from the sub-fascia.
Nivelar frente a los sub-fascia.
nee-beh-LAHR FREHN-teh ah lohs sub-fascia

Mark the wall.
Marcar la pared.
mahr-KAHR lah pah-REHD

Snap a line.
Marcar una línea con tiza.
mahr-KAHR OO-nah LEEN-yah kohn TEE-sah

Installing ledger

Nail the ledger to the house.
Clavar el travesaño a la casa.
klah-BAHR ehl trah-beh-SAH-nyoh ah lah KAH-sah

Figure 10. Soffit Framing
Estructura del plafón (sofito) exterior

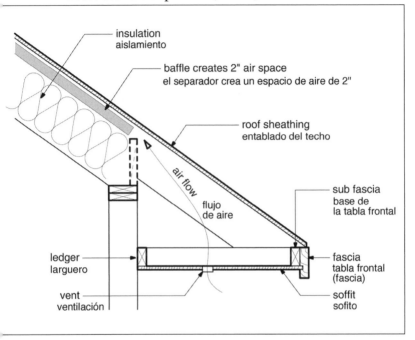

insulation
aislamiento

baffle creates 2" air space
el separador crea un espacio de aire de 2"

roof sheathing
entablado del techo

air flow
flujo
de aire

sub fascia
base de
la tabla frontal

ledger
larguero

fascia
tabla frontal
(fascia)

vent
ventilación

soffit
sofito

offit framing showing baffles installed to ventilate rafter bays.

Put the bottom on the line.
Poner la parte inferior en la línea.
poh-NEHR lah PAHR-teh een-fehr-YOHR ehn lah LEEN-yah

Installing blocks / joists

Install (the blocks / the joists).
Instalar (los bloques / las vigas).
een-stah-LAHR (lohs BLOH-kehs / lahs BEE-gahs)

Nail joists to each rafter.
Clavar las vigas a cada cabrio.
klah-BAHR lahs BEE-gahs ah KAH-dah KAH-bree-yoh

Soffit material (T&G)

Use this for the soffit.
Usar ésta para el sofito.
oo-SAHR EH-stah PAH-rah ehl soh-FEE-toh

Use one by (four / six / eight) tongue and groove.
Usar una lengüeta y ranura de uno por (cuatro / seis / ocho).
oo-SAHR OO-nah lehn-GWEH-tah ee rah-NOO-rah deh OO-noh pohr (KWAH-troh / sais / OH-choh)

Soffit material (plywood)

Use (three-eighths) inch plywood.
Usar madera tercida de (tres octavos) de pulgada.
oo-SAHR mah-DEH-ah tehr-see-AH-dah deh (trehs ohk-TAH-bohs) deh pool-GAH-dah

Installing soffit material (T&G)

Start at the fascia.
Comenzar con los fascia.
koh-mehn-SAHR kohn lohs fascia

Start with a full board.
Comenzar con una tabla entera.
koh-mehn-SAHR kohn OO-nah TAH-blah ehn-TEH-rah

Ripping the first board

Rip the first one to (four) inches.
Cortar el primero de unas (cuatro) pulgadas.
kohr-TAHR ehl pree-MEH-roh deh OO-nahs (KWAH-troh) pool-GAH-dahs

Rough / smooth side down

Put the (smooth / rough) side down.
Poner el lado (liso / áspero) hacia abajo.
poh-NEHR ehl LAH-doh (lee-SOH / ah-SPEH-roh) AH-syah ah-BAH-hoh

Soffit vents	**Install the soffit vent.** Instalar el respiradero del sofito. *een-stah-LAHR ehl rrreh-spee-rah-DEHR-oh* *dehl soh-FEE-toh*
	Put it (in the middle / against the house). Ponerlo (en el medio / contra la casa). *poh-NEHR-loh (ehn ehl MEH-dyoh /* *KOHN-trah lah KAH-sah)*
Nail guns	**Use (the finish gun / the staple gun).** Usar (la máquina de acabado / la engrapadora). *oo-SAHR (lah mah-KEE-nah deh ah-kah-BAH-doh /* *lah ehn-grah-pah-DOH-rah)*
	Use (the T-nailer / the siding gun). Usar (la clavadora en T / la pistola para revestimiento). *oo-SAHR (lah* klah-bah-DOH-rah ehn tee / lah pee-STOH-lah PAH-rah)
Nailing	**(Face / blind) nail it.** Clavarlo (de frente / cegado). *klah-BAHR-loh (deh FREH-teh / seh-GAH-doh)*
	Nail it every (six / eight) inches. Clavarlo cada (seis / ocho) pulgadas. *klah-BAHR-loh KAH-dah (sais / OH-choh) pool-GAH-dahs*
Nails	**Use galvanized nails.** Usar clavos galvanizados. *oo-SAHR KLAH-bohs gahl-bahn-nee-SAH-dohs*
	Do you need more nails? ¿Necesita más clavos? *neh-seh-SEE-tah mahs KLAH-bohs*
	The nails are in (my truck / the box). Los clavos están en (mi camión / la caja). *lohs KLAH-bohs EH-stahn ehn* *(mee kahm-YOHN / lah KAH-hah)*

SOFFITS (INTERIOR)

Soffit locations

Put a soffit here.
Poner un sofito aquí.
poh-NEHR oon soh-FEE-toh ah-KEE

It (starts / stops) here.
(Comienza / termina) aquí.
(kohm-YEHN-sah / tehr-MEE-nah) ah-KEE

It goes around the edge of the room.
Recorre todo el borde del cuarto.
rrreh-KOH-rrreh TOH-doh ehl BOHR-deh dehl KWAHR-toh

Soffit height

Make it (ninety-seven) inches to the bottom.
Que esté a (noventa y siete) pulgadas del piso.
keh EH-steh ah (noh-BEHN-tah ee SYEH-teh) pool-GAH-dahs dehl PEE-soh

It steps (up / down) here.
(Sube / baja) aquí.
(SOO-beh / BAH-hah) ah-KEE

Material size

Use two by (four / six).
Usar dos por (cuatro / seis).
oo-SAHR dohs pohr (KWAH-troh / sais)

Joist direction

The joists run this way.
Las vigas van en esta dirección.
lahs BEE-gahs bahn ehn EH-stah dee-rehk-SYOHN

Pulling measurements

(Butt / hook) the wall.
(Tocar / enganchar) la pared.
toh-KAHR / ehn-gahn-CHAHR) lah pah-REHD

Mark it at (fifty) inches.
Marcarla a (cincuenta) pulgadas.
mahr-KAHR-lah ah (seen-KWEHN-tah) pool-GAH-dahs

Soffit layout

Mark (the corners / the ceiling).
Marcar (las esquinas / el techo).
mahr-KAHR (lahs eh-SKEE-nahs / ehl TEH-choh)

Snap lines from corner to corner.
Marcar con tiza las líneas de esquina a esquina.
mar-KAHR kohn TEE-sah lahs LEEN-yahs deh
eh-SKEE-nah ah eh-SKEE-nah

Snap a line on the ceiling.
Marcar una línea con tiza en el techo.
mar-KAHR OO-nah LEEN-yah kohn
TEE-sah ehn ehl TEH-choh

Installing the ledger

Put a ledger on the wall.
Poner un ledger en la pared.
poh-NEHR oon ledger ehn lah pah-REHD

Put it above the line.
Ponerlo sobre la línea.
poh-NEHR-loh SOH-breh lah LEEN-yah

Installing verticals

Install the verticals on the line.
Instalar los verticales en la línea.
een-stah-LAHR lohs behr-tee-KAH-lehs ehn lah LEEN-yah

Nail two by (four) to the joists.
Clavar dos por (cuatro) en las vigas.
klah-BAHR dohs pohr (KWAH-troh) ehn lahs BEE-gahs

(Eleven) inches to the bottom.
A (once) pulgadas del piso.
ah (OHN-seh) pool-GAH-dahs dehl PEE-soh

Nail a horizontal to them.
Clavar una horizontal en ellas.
klah-BAHR OO-nah oh-ree-sohn-TAHL ehn EH-yahs

Joist layout

Mark the joist layout.
Marcar cada lugar donde se colocarán las vigas.
mahr-KAHR KAH-dah loo-GAHR DOHN-deh
seh koh-loh-kah-RAHN lahs BEE-gahs

Put them on (sixteen / twenty-four)-inch centers.
Centrarlas a (dieciséis / veinticuatro) pulgadas.
sehn-TRAHR-lahs ah (dyehs-EE-sais /
behn-tee-KWAH-troh) pool-GAH-dahs

Special layout	**Center a bay here.**
	Centrar un vano aquí.
	sehn-TRAHR oon BAH-noh ah-KEE
	(Fifty) inches to center.
	A (cincuenta) pulgadas del centro.
	ah (seen-KWEHN-tah) pool-GAH-dahs dehl SEHN-troh
	Layout is marked on the floor.
	La disposición está marcada en el piso.
	lah dees-poh-see-SYOHN EH-stah
	mahr-KAH-dah ehn ehl PEE-soh
Cutting pressure blocks	**Cut (twenty) blocks.**
	Cortar (veinte) bloques.
	kohr-TAHR (BAIN-teh) BLOH-kehs
	Cut them (fourteen / twenty-two) and one-half inches.
	Cortarlos de (catorce / veintidós) y media pulgada.
	kohr-TAHR-lohs deh (kah-TOHR-seh / behn-tee-DOHS)
	ee MEH-dyah pool-GAH-dah
Installing pressure blocks	**Install the pressure blocks.**
	Instalar los bloques de presión.
	een-stah-LAHR lohs BLOH-kehs deh preh-SYOHN
	Put one in every other bay.
	Poner uno en cada otro vano.
	poh-NEHR OO-noh ehn KAH-dah OH-troh BAH-noh
Cutting joists	**Cut them all the same length.**
	Cortarlas de la misma longitud.
	kohr-TAHR-lahs deh lah MEES-mah lohn-hee-TOOD
	Cut them (thirty-eight and one-quarter).
	Cortarlas de (treinta y ocho y un cuarto).
	kohr-TAHR-lahs deh (TRAIN-tah
	ee OH-choh ee oon KWAHR-toh)
Installing joists	**Install the joists.**
	Instalar las vigas.
	een-stah-LAHR lahs BEE-gahs

Install them (flat / on edge).
Instalarlas sobre (el lado plano / el borde).
een-stah-LAHR-lahs SOH-breh
(ehl LAH-doh PLAH-noh / ehl BOHR-deh)

Backing

Put backing here.
Poner el respaldo aquí.
poh-NEHR ehl rrreh-SPAHL-doh ah-KEE

Firestop (blocks)

Put firestop blocks here.
Poner los bloques de antifuego aquí.
poh-NEHR lohs BLOH-kehs deh ahn-tee-FWEH-goh ah-KEE

Put a row of blocks here.
Poner una hilera de bloques aquí.
poh-NEHR OO-nah ee-LEH-rah deh BLOH-kehs ah-KEE

Chapter 7 ROOFING

ASPHALT AND WOOD SHINGLES

Installing drip edge

Install the drip edge.
Instalar el borde de goteo.
een-stah-LAHR ehl BOHR-deh deh goh-TEH-oh

Put it where the roof is (level / sloped).
Ponerlo donde el techo está (nivelado / inclinado).
*poh-NEHR-loh DOHN-deh ehl TEH-choh EH-stah
(nee-beh-LAH-doh / een-klee-NAH-doh)*

Nail it every (eighteen) inches.
Clavarlo cada (dieciocho) pulgadas.
*KLAH-bahr-loh KAH-dah (dyehs-ee-OH-choh)
pool-GAH-dahs*

Installing valley flashing (metal)

Install the valley flashing.
Instalar el vierteaguas de hoyas.
een-stah-LAHR ehl behr-teh-AH-gwahs deh HOH-yahs

Start at the bottom.
Comenzar abajo.
koh-mehn-SAHR ah-BAH-hoh

Nailing valley flashing

Nail the edges only.
Clavar sólo los bordes.
klah-BAHR SOH-loh lohs BOHR-dehs

Installing valley flashing (paper)

Put paper in the valley.
Poner papel en la hoya.
poh-NEHR pah-PEHL ehn lah HOH-yah

Use the (thirty / ninety) pound paper.
Usar papel de (treinta / noventa) libras.
*oo-SAHR pah-PEHL deh (TRAIN-tah /
noh-BEHN-tah) LEE-brahs*

It must touch the roof in the center.
Debe tocar el techo en el centro.
DEH-beh toh-KAHR ehl TEH-choh ehn ehl SEHN-troh

Staple the edges only.
Grapar sólo los bordes.
grah-PAHR SOH-loh lohs BOHR-dehs

Installing chimney flashing

Install the chimney flashing.
Instalar el vierteaguas de la chimenea.
een-stah-LAHR ehl byehr-teh-AH-gwahs deh lah chee-meh-NEH-ah

Use the small pieces.
Usar las piezas pequeñas.
oo-SAHR lahs PYEH-sahs peh-KEHN-yahs

One piece for each shingle.
Una pieza para cada teja.
OO-nah PYEH-sah PAH-rah KAH-dah TEH-hah

UNDERLAYMENT

Installing paper

Install the paper.
Instalar el papel.
een-stah-LAHR ehl pah-PEHL

Install it (with / without) tar.
Instalarlo (con / sin) alquitrán.
een-stah-LAHR-lo (kohn / seen) ahl-kee-TRAHN

It must (flat / smooth).
Debe quedar (plano / liso).
DEH-beh keh-DAHR (PLAH-noh / LEE-soh).

Use the (thirty / ninety) pound paper.
Usar el papel de (treinta / noventa) libras.
oo-SAHR ehl pah-PEHL de (TRAIN-tah / noh-BEHN-tah) LEE-brahs

Start at the bottom.
Empezar abajo.
ehm-peh-SAHR ah-BAH-hoh

Overlaps

Overlap each course (two) inches.
Traslapar cada hilada (dos) pulgadas.
trahs-lah-PAHR KAH-dah ee-LAH-dah (dohs) pool-GAH-dahs

Overlap the ends (three) inches.
Traslapar los extremos (tres) pulgadas.
trahs-lah-PAHR lohs ehk-TREH-mohs (trehs) pool-GAH-dahs

It must overlap the course below (three) inches.
Debe haber un traslapo de (tres) pulgadas
encima de la hilada por debajo.
DEH-beh ah-BEHR oon trahs-LAH-poh deh (trehs) pool-GAH-dahs ehn-SEE-mah deh-lah ee-LAH-dah pohr deh-BAH-hoh

Staggering breaks

Stagger the breaks.
Extender los puntos de unión de los extremos.
ehk-TEHN-dehr lohs POON-tohs deh oo-NYON deh lohs ehk-TREH-mohs

Spreading tar

Spread the tar.
Extender el alquitrán.
ehk-TEHN-dehr ehl ahl-kee-TRAHN

Cover the paper completely.
Cubrir el papel completamente.
koo-BREER ehl pah-PEHL kohm-pleh-tah-MEHN-teh

Put more here.
Poner más aquí.
poh-NEHR mahs ah-KEE

Paper exposure

Leave (nine / twelve / eighteen) inches exposed.
Dejar (nueve / doce / dieciocho) pulgadas expuestas.
deh-HAHR (NWEH-beh / DOH-seh / dyehs-ee-OH-choh) pool-GAH-dahs ehks-PWEH-stahs

Remove the air bubbles.
Remover las burbujas de aire.
rrreh-moh-BEHR lahs boor-BOO-hahs deh ah-EE-reh

Number of layers	**Install (two / three / four) layers of paper.**
	Instalar (dos / tres/ cuatro) capas de papel.
	een-stah-LAHR (dohs / trehs / KWAH-troh)
	KAH-pahs deh pah-PEHL
Spreading gravel	**Spread the gravel.**
	Esparcir la grava.
	ehs-pahr-SEER lah GRAH-bah
	It must cover the roof completely.
	Debe cubrir el techo completamente.
	DEH-beh koo-BREER ehl TEH-choh
	kohm-pleh-tah-MEHN-teh
Stapler	**Use the stapler.**
	Usar la engrapadora.
	oo-SAHR lah ehn-grah-pah-DOHR-ah
	Staple it every (sixteen) inches.
	Graparlo cada (dieciséis) pulgadas.
	grah-PAHR-loh KAH-dah (dyehs-ee-SAIS) pool-GAH-dahs
Nailing felt	**Nail it every (twelve) inches.**
	Clavarlo cada (doce) pulgadas.
	klah-BAHR-loh KAH-dah (DOH-seh) pool-GAH-dahs
	...at the edge / in the middle
	...en el borde / en el medio
	...ehn ehl BOHR-deh / ehn ehl MEH-dyoh
Nails	**Use (one and one-quarter)-inch nails.**
	Usar clavos de (una y un cuarto de) pulgada.
	oo-SAHR KLAH-bohs deh (OO-nah ee oon
	KWAHR-toh deh) pool-GAH-dah

■ SHINGLE LAYOUT AND INSTALLATION

Pulling layout	**Mark layout on the roof.**
	Haga el trazo en el techo.
	AH-gah ehl TRAH-soh ehn ehl TEH-choh

Mark it every (fifteen / thirty) inches.
Marcarlo cada (quince / treinta) pulgadas.
mar-KAHR-loh KAH-dah (KEEN-seh /
TRAIN-tah) pool-GAH-dahs

…from the bottom
…desde abajo
…DEHS-deh ah-BAH-hoh

Snapping lines	**Snap the lines.** Marcar las líneas con tiza. *mahr-KAHR las LEEN-yahs kohn TEE-sah*
Installing shingles	**Install the shingles.** Instalar las tejas. *een-stah-LAHR lahs TEH-hahs* **Start at the bottom.** Empezar abajo. *ehm-peh-SAHR ah-BAH-hoh*
Doubling the first course	**The first course is two layers.** La primera hilada es de dos capas. *Lah pree-MEH-rah ee-LAH-dah ehs deh dohs KAH-pahs*
Reversing the shingle	**On the first layer, put the tabs up.** En la primer zcapa, poner las lengüetas hacia arriba. *ehn ehl pree-MEH-rah KAH-pah, poh-NEHR lahs* *lehn-GWEH-tahs AH-syah ah-RRREE-bah*
Shingle overhang	**It must overhang the edge (one and one-half) inch(es).** Debe sobrepasar el borde en (una y media) pulgada(s) *DEH-beh soh-breh-pah-SAHR ehl BOHR-deh ehn* *(OO-nah ee MEH-dyah) pool-GAH-dah(s)* **…at the (bottom / sides)** …(abajo / a los lados) *…(ah-BAH-hoh / ah lohs LAH-dohs)*
Shingle exposure	**Leave (five and one-half) inches exposed.** Dejar (cinco y media) pulgadas expuesta(s). *deh-HAHR (SEEN-koh ee MEH-dyah)* *pool-GAH-dahs ehks-PWEH-stahs*

Figure 11. Asphalt Shingle Application
Instalación de las teja asfáltica

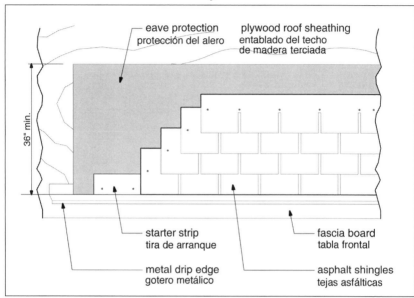

Starter course, underlayment, and drip edge details for asphalt shingles.

Figure 12. Wood Shingle Application
Instalación de tejas de madera

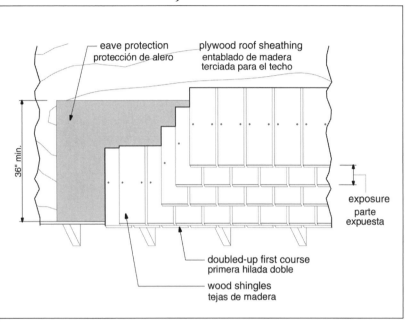

Starter course details, nailing schedules, and exposure table for wood shingles.

Shingle spacing	**Leave (one quarter / three eighths) inch between them.** Dejar (un cuarto / tres octavos) de pulgada entre ellas. *deh-HAHR (oon KWAHR-toh / trehs ohk-TAH-bohs)* *pool-GAH-dah EHN-treh EH-yahs*
Offsetting breaks	**Offset the breaks (three) inches.** Compensar (tres) pulgadas en los puntos de unión de los extremos. *kohm-pehn-SAHR (trehs) pool-GAH-dahs ehn lohs* *POON-tohs deh oo-NYON deh lohs ehk-TREH-mohs*
	The breaks must not align. Los puntos de unión de los extremos no deben alinearse. *Lohs POON-tohs deh oo-NYON deh lohs* *ehk-TREH-mohs noh DEH-beh ah-lee-neh-AHR-seh*
	…in (three) courses …en (tres) hiladas *…ehn (trehs) ee-LAH-dahs*
Nailing schedule	**Put nails here.** Poner clavos aquí. *poh-NEHR KLAH-bohs ah-KEE*
	Use (two / six) nails per shingle. Usar (dos / seis) clavos por teja. *oo-SAHR (dohs / sais) KLAH-bohs pohr TEH-hah*
Nails	**Use (one and one-quarter)-inch nails.** Usar clavos de (una y un cuarto de) pulgada. *oo-SAHR KLAH-bohs deh (OO-nah ee oon* *KWAHR-toh deh) pool-GAH-dah*
Flashing skylights	**Install flashing at the skylight.** Instalar el vierteaguas en el tragaluz. *een-stah-LAHR ehl byehr-teh-AH-gwahs* *ehn ehl trah-gah-LOOS*
	Put the flashing (over / under) the shingles here. Poner el vierteaguas (sobre / debajo) de las tejas aquí. *poh-NEHR ehl byehr-teh-AH-gwahs (SOH-breh /* *de-BAH-hoh) deh lahs TEH-hahs ah-KEE*

Installing stack flashing	**Install flashing at the pipes.** Instalar el vierteaguas en las cañerías. *een-stah-LAHR ehl byehr-teh-AH-gwahs* *ehn lahs kahn-YEHR-yahs*

Watch, do it like this.
Mire, así.
MEE-reh, ah-SEE

Roof to wall junctions (step-flashing)	**Install flashing at the wall.** Instalar el vierteaguas en la pared. *een-stah-LAHR ehl byehr-teh-AH-gwahs ehn lah pah-REHD*

Use one piece per shingle.
Usar una pieza por teja.
oo-SAHR OO-nah PYEH-sah pohr TEH-hah

VALLEY INSTALLATION

Open valley	**This roof gets open valleys.** Este techo tiene hoyas abiertas. *EH-steh TEH-choh TYEH-neh HOH-yahs ah-BYEHR-tahs*

Stop the shingles here.
Pare las tejas aquí.
PAH-reh lahs TEH-hahs ah-KEE

(Three) inches from the center of the valley.
A (tres) pulgadas del centro de la hoya.
ah (trehs) pool-GAH-dahs dehl SEHN-troh deh lah HOH-yah

...at the (top / bottom)
...(arriba / abajo)
...(ah-RRREE-bah / ah-BAH-hoh).

Closed valley (asphalt shingle)	**This roof gets closed valleys.** Este techo tiene hoyas cerradas. *EH-steh TEH-choh TYEH-neh HOH-yahs seh-RRRAH-dahs*

Run this side through.
Atravesar este lado.
ah-trah-beh-SAHR EH-steh LAH-doh

This side stops at the valley.
Este lado para en la hoya.
EH-steh LAH-doh PAH-rah ehn lah HOH-yah

Woven valley
(asphalt shingle)

This roof gets woven valleys.
Este techo tiene hoyas tejidas.
EH-steh TEH-choh TYEH-neh HOH-yahs teh-HEE-dahs

Install both sides at the same time.
Instalar ambos lados al mismo tiempo.
een-stah-LAHR AHM-bohs LAH-dohs ahl
MEES-moh TYEHM-poh

Both sides run through.
Ambos lados atraviesan.
AHM-bohs LAH-dohs ah-trah-bee-eh-SAHN

RIDGE VENT AND SHINGLE INSTALLATION

Ridge vents

Do you know how to install a ridge vent?
¿Sabe cómo se instala un respiradero de caballete?
SAH-beh KOH-moh seh ihn-STAH-lah
oon deh kah-bah-YET-teh

Put the ridge vent here.
Poner la ventilación de caballete aquí.
poh-NEHR lah behn-tee-lah-SYOHN
deh kah-bah-YEH-teh ah-KEE

Put the foam here.
Poner la espuma aquí.
poh-NEHR lah eh-SPOO-mah ah-KEE

Ridge shingles

Install the shingles at the ridge.
Instalar las tejas en el caballete.
een-stah-LAHR lahs TEH-hahs ehn ehl kah-bah-YEH-teh

Use (two)-inch nails.
Usar clavos de (dos) pulgadas.
oo-SAHR KLAH-bohs deh (dohs) pool-GAH-dahs

Don't compress the foam.
No comprimir la espuma.
noh kohm-pree-MEER lah eh-SPOO-mah

◼ COMPRESSOR, HOSES, AND NAIL GUN

**Setting up the
compressor**

Set up the compressor.
Alistar el compresor.
ah-lees-TAHR ehl kohm-preh-SOHR

It must be level.
Debe quedar nivelado.
DEH-beh keh-DAHR nee-beh-LAH-doh

Hoses

Roll out the hoses.
Desenrollar las mangueras.
dehs-ehn-rrroh-YAHR lahs mahn-GEH-rahs

Nail gun

Use the nail gun.
Usar la clavadora.
oo-SAHR lah klah-bah-DOH-rah

◼ INTERLAYMENT

**Installing the
interlayment**

Use the (two) foot rolls of paper.
Usar los rollos de papel de (dos) pies.
oo-SAHR lohs RRROH-yohs deh pah-PEHL deh (dohs) pyehs

Put paper between each course of shakes.
Poner papel entre cada hilada de tejamaniles.
poh-NEHR pah-PEHL EHN-treh
KAH-dah ee-AH-dah deh teh-ha-mah-NEE-lehs

Interlayment overlap

Overlap the shakes (three / four) inches.
Traslapar los tejamaniles (tres / cuatro) pulgadas.
trahs-lah-PAHR lohs teh-ha-mah-NEE-lehs
(trehs / KWAH-troh) pool-GAH-dahs

Figure 13. Wood Shake Installation
Colocación de tejamanil

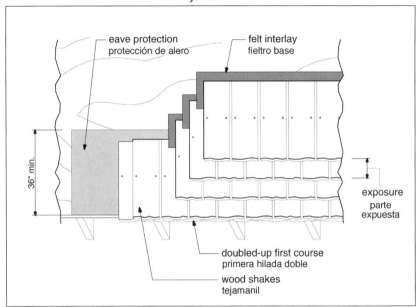

eave protection
protección de alero

felt interlay
fieltro base

36" min.

exposure
parte
expuesta

doubled-up first course
primera hilada doble

wood shakes
tejamanil

Wood shake installation with starter course and interlayment details and exposure table.

verhang

They must overhang (one and one-half) inch(es).
Deben sobrecolgar (una y media) pulgada(s).
DEH-behn soh-breh-kohl-GAHR
(OO-nah ee MEH-dyah) pool-GAH-dah(s)

...at the (bottom / sides)
...(abajo / a los lados)
...(ah-BAH-hoh / ah lohs LAH-dohs)

CONCRETE TILE

atten layout

Lay out for the battens.
Hacer el trazo para los listones.
ah-SEHR ehl TRA-soh PAH-rah lohs lee-STOH-nehs

Snap lines every (ten) inches.
Marcar las líneas con tiza cada (diez) pulgadas.
mahr-KAHR las LEEN-yahs kohn TEE-sah
KAH-dah (dyehs) pool-GAH-dahs

...from the bottom edge
...desde el borde inferior
...DEHS-deh ehl BOHR-deh een-fehr-YOHR

stalling battens

Install the battens.
Instalar los listones.
een-stah-LAHR lohs lee-STOH-nehs

**atten nailing
chedule**

Nail them every (twelve) inches.
Clavarlos cada (doce) pulgadas.
klah-BAHR-lohs KAH-dah (DOH-seh) pool-GAH-dahs

stalling tiles

Install the tiles.
Instalar las tejas.
een-stah-LAHR lahs TEH-hahs

Watch, do it like this.
Mire, así.
MEE-reh, ah-SEE

Cutting tiles	**Cut the tile here.** Cortar la teja aquí. *kohr-TAHR lah TEH-hah ah-KEE* **Do you need a (saw / blade)?** ¿Necesita una (sierra / cuchilla)? *neh-seh-SEE-tah OO-nah (SYEH-rrrah / koo-CHEE-yah)*

▮▮▮▮▮▮ CLAY TILE

Hip backing	**Nail two by (four / six) on the hips.** Clavar dos por (cuatro / seis) en las limas tesas. *klah-BAHR dohs pohr (KWAH-troh / sais)* *ehn lahs LEE-mahs TEH-sahs* **Put them on edge.** Ponerlas en el borde. *poh-NEHR-lahs ehn ehl BOHR-deh*
Vertical batten layout	**Lay out for the battens.** Hacer el trazo para los listones. *ah-SEHR ehl TRA-soh pah-rah lohs lee-STOH-nehs* **They go vertically.** Van verticalmente. *bahn behr-tee-kahl-MEHN-teh* **Put one every (eight) inches.** Poner uno cada (ocho) pulgadas. *poh-NEHR OO-noh KAH-dah (OH-choh) pool-GAH-dahs* **Start in the center of the roof.** Empezar en el centro del techo. *ehm-peh-SAHR ehn ehl SEHN-troh dehl TEH-choh*
Installing tiles	**Install the tiles.** Instalar las tejas. *een-stah-LAHR lahs TEH-hahs*
Tile exposure	**Leave (ten) inches exposed.** Dejar (diez) pulgadas expuestas. *deh-HAHR (dyehs) pool-GAH-dahs ehks-PWEH-stahs*

Put mortar here.
Poner mortero aquí.
poh-NEHR mohr-TEH-roh ah-KEE

BUILT-UP ROOFING

Set up the kettle here.
Instalar el crisol aquí.
een-stah-LAHR ehl kree-SOHL ah-KEE

Start the kettle.
Prender el crisol.
prehn-DEHR ehl kree-SOHL

Is it ready?
¿Está listo?
EH-stah LEE-stoh

Set up the hose.
Preparar la manguera.
preh-pah-RAHR lah mahn-GEH-rah

Install the cant strips.
Instalar los listones chaflanados.
een-stah-LAHR lohs lee-STOH-nehs

GUTTER AND DOWNSPOUTS

Install the gutter.
Instalar los canales.
een-stah-LAHR lohs kah-NAH-lehs

It must slope down this way.
Debe inclinarse hacia abajo así.
DEH-beh een-klah-NAHR-seh AH-syah ah-BAH-hoh ah-SEE

Put a downspout here (and there).
Poner un bajante aquí (y allá).
poh-NEHR oon bah-HAHN-teh ah-KEE (ee ah-YAH)

Figure 14. Gutter Section
Sección transversal del canalón

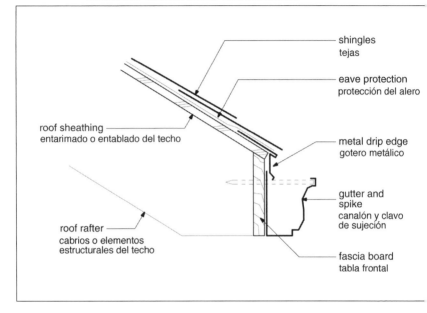

shingles
tejas

eave protection
protección del alero

roof sheathing
entarimado o entablado del techo

metal drip edge
gotero metálico

gutter and
spike
canalón y clavo
de sujeción

roof rafter
cabrios o elementos
estructurales del techo

fascia board
tabla frontal

Chapter 8 INSULATION

nloading

Unload the truck.
Descargar el camión.
dehs-kahr-GAHR ehl kahm-YOHN

tocking the house

Stock the (house / building).
Cargar (la casa / el edificio).
kahr-GAHR (lah KAH-sah / ehl eh-dee-FEE-syoh)

ocation

Put the (R-nineteen / R-thirty) here.
Poner el (R-diecinueve / R-treinta) aquí.
*poh-NEHR ehl (R-dyehs-ee-NWEH-beh /
R-TRAIN-tah) ah-KEE*

...on (the first / the second / the third) floor
...en el (primer / segundo / tercer) piso
...ehn ehl (pree-MEHR / seh-GOON-doh / tehr-SEHR) PEE-soh

This is for the (walls / ceiling / attic / crawlspace).
Este es para (las paredes / el cielo raso / el ático / el gatero).
*EH-steh ehs PAH-rah (lah pah-REHD-ehs / ehl SYEH-loh
RRRAH-soh / ehl AH-tee-koh / ehl gah-TEH-roh)*

/pe

Put (fiberglass / blown-in) here.
Poner (fibra de vidrio / soplado) aquí.
*poh-NEHR (FEE-brah deh bee-DREE-oh /
soh-PLAH-doh) ah-KEE*

INSTALLATION: FIBERGLASS BATT

Put fiberglass batt here.
Poner el batt de fibra de vidrio aquí.
poh-NEHR ehl batt de FEE-brah deh bee-DREE-oh ah-KEE

eiling Installation

Insulate the ceiling.
Aislar el cielo raso.
ah-ees-LAHR ehl SYEH-loh RRRAH-soh

Cleaning the floor	**Clean the floor.** Limpiar el piso. *leem-PYAHR ehl PEE-soh*
Scaffolding	**Set up the scaffolding.** Armar el andamiaje. *ahr-MAHR ehl ahn-dah-MYAH-heh*
Installing baffles	**Install the baffles.** Instalar las placas de desviación. *een-stah-LAHR lahs PLAH-kahs deh dehs-bee-ah-SYON* **There must be a two-inch air space.** Debe quedar un espacio de aire de dos pulgadas. *DEH-beh keh-DAHR oon eh-SPAH-syoh* *deh ah-EE-reh deh dohs pool-GAH-dahs*
Light fixtures	**Don't cover the light fixture.** No cubrir el artefacto de la luz. *no koo-BREER ehl ahr-teh-FAHK-toh deh lah loos*
Insulation specifications	**This insulation is (twenty-three) inches wide.** Este aislamiento tiene (veintitrés) pulgadas de ancho. *EH-steh ah-ees-lah-MYEHN-toh TYEH-neh* *(behn-tee-TREHS) pool-GAH-dahs deh AHN-choh* **This insulation is (R-thirty).** Este aislamiento es (R-treinta). *EH-steh ah-ees-lah-MYEHN-toh ehs (R-TRAIN-tah)*
Wall installation	**Insulate the walls.** Aislar las paredes. *ah-ees-LAHR lah pah-REHD-ehs*
Insulation specifications	**This insulation is (fifteen) inches wide.** Este aislamiento tiene (quince) pulgadas de ancho. *EH-steh ah-ees-lah-MYEHN-toh TYEH-neh* *(KEEN-seh) pool-GAH-dahs deh AHN-choh* **This insulation is (R-nineteen).** Este aislamiento es (R-diecinueve). *EH-steh ah-ees-lah-MYEHN-toh ehs* *(R-dyehs-ee-NWEH-beh)*

Figure 15. Installing Baffles
Instalación de las placas de desviación

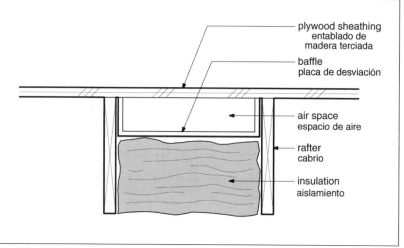

plywood sheathing
entablado de
madera terciada

baffle
placa de desviación

air space
espacio de aire

rafter
cabrio

insulation
aislamiento

Baffles create an air space to ventilate rafter bays in vaulted ceilings.

Installing insulation	**It must fit tightly.** Debe quedar apretado. *DEH-beh keh-DAHR ah-preh-TAH-doh* **Cut around the boxes.** Cortar alrededor de las cajas. *kohr-TAHR ahl-rrreh-deh-DOR deh lahs KAH-hahs* **We need more here.** Necesitamos más aquí. *neh-seh-see-TAH-mohs mahs ah-KEE*
Insulation orientation	**The (paper / foil) side goes in.** El lado de (papel / lámina) va hacia adentro. *ehl LAH-doh deh (pah-PEHL /* *LAH-mee-nah) bah AH-syah ah-DEHN-troh*
Stapling	**Do you have a stapler?** ¿Tiene una engrapadora? *TYEH-neh OO-nah ehn-graph-pah-DOH-rah* **Do you need staples?** ¿Necesita grapas? *neh-seh-SEE-tah GRAH-pahs*
Insulating crawlspaces	**Insulate the crawlspace.** Aislar el gatero. *ah-ees-LAHR ehl gah-TEH-roh* **Insulate the floor.** Aislar el piso. *ah-ees-LAHR ehl PEE-soh*
Insulation type	**Use (R-thirty) insulation.** Usar aislamiento (R-treinta). *oo-SAHR ah-ees-lah-MYEHN-toh (R-TRAIN-tah)*
Insulation orientation	**The (paper / foil) side goes up.** El lado de (papel / lámina) va hacia arriba. *ehl LAH-doh deh (pah-PEHL / LAH-mee-nah)* *bah AH-syah ah-RRREE-bah*

Watch, do it like this.
Mire, así.
MEE-reh, ah-SEE

Method of securing

Keep it in place with wire.
Mantenerlo en su lugar con alambre.
mahn-teh-NEHR-loh ehn soo loo-GAHR kohn ah-LAHM-breh

Tack nails in the joists.
Fijar clavos en las vigas.
fee-HAHR KLAH-bohs ehn lahs BEE-gahs

Stretch wire across the joists.
Estirar alambre sobre las vigas.
ehs-tee-RAHR ah-LAHM-breh SOH-breh lahs BEE-gahs

Wind the wire around the nails.
Enredar el alambre alrededor de los clavos.
ehn-reh-DAHR ehl ah-LAHM-breh
ahl-rrreh-deh-DOR deh lohs KLAH-bohs

Insulating pipes

Wrap the (copper / galvanized) pipes with insulation.
Envolver la tubería (de cobre / galvanizada) con aislamiento.
ehn-bohl-BEHR lah too-beh-REE-ah (deh KOH-breh /
gahl-bah-nee-SAH-dah) con ah-ees-lah-MYEHN-toh

They must be completely covered.
Debe quedar completamente cubierta.
DEH-beh keh-DAHR kohm-pleh-tah-MEHN-teh
koo-BYEHR-tah

Foam insulation

Use the foam insulation.
Usar el aislamiento de espuma.
oo-SAHR ehl ah-ees-lah-MYEHN-toh deh eh-SPOO-mah

Hold it in place with tape.
Mantenerlo en su lugar con cinta.
mahn-teh-NEHR-loh ehn soo loo-GAHR con SEEN-tah

Insulating heating ducts

Insulate the heating ducts.
Aislar los tubos de calefacción.
ah-ees-LAHR lohs TOO-bohs deh kah-leh-fahk-SYOHN

Wrap the heating ducts with insulation.
Envolver los tubos de calefacción con aislamiento.
ehn-bohl-BEHR lohs TOO-bohs deh
kah-leh-fahk-SYOHN kohn ah-ees-lah-MYEHN-toh

Hold it in place with tape.
Mantenerlo en su lugar con cinta.
mahn-teh-NEHR-loh ehn soo loo-GAHR con SEEN-tah

Insulating dryer / bathroom vents

Insulate the (dryer / bathroom) vents.
Aislar las ventilaciones de (la secadora / el baño).
ah-ees-LAHR lahs behn-tee-lah-SYOH-nehs
deh (lah seh-kah-DOH-rah / ehl BAHN-yoh)

Wrap this duct with insulation.
Envolver este tubo con aislamiento.
ehn-bohl-BEHR EH-steh TOO-boh
kohn ah-ees-lah-MYEHN-toh

■■■■■■■■■ MOISTURE BARRIER

Installing plastic

Install the plastic.
Instalar el plástico.
een-stah-LAHR ehl PLAH-stee-koh

It must fit into the corners.
Debe entrar en las esquinas.
DEH-beh ehn-TRAHR ehn lahs eh-SKEE-nahs

Insulation: blown-in

Put blown-in insulation here.
Poner el aislamiento soplado aquí.
poh-NEHR ehl ah-ees-lah-MYEHN-toh soh-PLAH-doh ah-KEE

■■■■■■■■■ SET-UP AND ACCESS

Truck set-up

Park the truck (here / there).
Parquear el camión (aquí / allá).
pahr-keh-AHR ehl kahm-YOHN (ah-KEE / ah-YAH)

Run the hose through the (door / window).
Pasar la manguera por la (puerta / ventana).
pah-SAHR lah mahn-GEH-rah pohr
lah (PWEHR-tah / behn-TAH-nah)

Protect the window.
Proteger la ventana.
proh-teh-HEHR lah behn-TAH-nah

Compressor set-up

Set up the compressor.
Armar el compresor.
ahr-MAHR ehl kom-preh-SOHR

Roll out the hoses.
Desenrollar las mangueras.
dehs-ehn-roh-YAHR lahs mahn-GEH-rahs

Connect the staple gun(s).
Conectar la(s) engrapadora(s).
koh-nehk-TAHR lah(s) ehn-grah-pah-DOH-rah(s)

▉ BLOWING ATTIC INSULATION

Blowing attics

Insulate the attic.
Aislar ático.
ah-ees-LAHR AH-tee-koh

It must be (fourteen) inches deep.
Debe tener (catorce) pulgadas de profundidad.
DEH-beh teh-NEHR (kah-TOHR-seh)
pool-GAH-dahs deh proh-foon-dee-DAHD

Access

Put the ladder here.
Poner la escalera aquí.
poh-NEHR lah eh-skah-LEH-rah ah-KEE

Go in through here.
Entrar por aquí.
ehn-TRAHR pohr ah-KEE

Avoiding ceiling damage	**Step on the framing only.**
	Pisar sólo el armazón.
	pee-SAHR SOH-loh ehl ahr-mah-SOHN
	Don't step on the (drywall / insulation).
	No pisar (la hoja de yeso / el aislamiento).
	noh pee-SAHR (lah OH-ah dehYEH-soh /
	ehl ah-ees-lah-MYEHN-toh)
	If you step on the ceiling it will break.
	Si pisa el cielo raso se romperá.
	see PEE-sah ehl SYEH-loh RRRAH-soh seh rrrohm-PEH-rah
Light fixtures	**Remove the insulation from above the light fixture.**
	Remover el aislamiento por encima del artefacto de la luz.
	rrreh-moh-BEHR ehl ah-ees-lah-MYEHN-toh
	pohr ehn-SEE-mah dehl ahr-teh-FAHK-toh deh lah loos
Attic vents	**Don't cover the attic vents.**
	No cubrir las ventilaciones del ático.
	noh koo-BREER lahs behn-tee-lah-SYOH-nehs
	dehl AH-tee-koh
	When you are finished, clear the attic vents.
	Al terminar, despejar las ventilaciones del ático.
	ahl tehr-mee-NAHR, dehs-peh-HAHR lahs
	behn-tee-lah-SYOH-nehs dehl AH-tee-koh

▬▬▬▬▬▬ BLOWING WALLS

Wall fabric	**(Install / staple) the fabric.**
	(Instalar / engrapar) el tejido.
	(een-stah-LAHR / ehn-grah-PAHR) ehl teh-HEE-doh
	It must be tight.
	Debe entrar apretado.
	DEH-beh ehn-TRAHR ah-preh-TAH-doh
	It must fit into the corners.
	Debe cuadrar en las esquinas.
	DEH-beh kwah-DRAHR ehn las eh-SKEE-nahs

Foaming doors and windows	**Foam the doors and windows.** Espumar las puertas y ventanas. *eh-spoo-MAHR lahs PWEHR-tahs ee behn-TAH-nahs*
Foam	**Use the minimally expanding foam.** Usar la espuma de mínima expansión. *oo-SAHR lah eh-SPOO-mah deh* *MEE-nee-mah ehk-spahn-SYOHN*
Areas to foam	**Put foam around the (door / window).** Poner espuma alrededor del larguero de la (puerta / ventana). *poh-NEHR eh-SPOO-mah ahl-rrreh-deh-DOHR* *deh lahr-GEH-roh deh lah (PWEHR-tah / behn-TAH-nah)*
Applying foam	**Keep it moving.** Mantenerla en movimiento. *mahn-tah-NEHR-lah ehn moh-bee-MYEHN-toh* **Don't use too much.** No usar demasiado. *noh oo-SAHR deh-mah-SYAH-doh*

Chapter 9   DRYWALL

STOCKING THE BUILDING

Loading and unloading drywall

(Load / unload) the truck.
(Cargar / descargar) el camión.
(kahr-GAHR / dehs-KAHR-gahr) ehl kahm-YOHN

Put the drywall in (the house / the truck).
Poner las hojas de yeso en (la casa / el camión).
poh-NEHR lahs OH-hahs deh YEH-soh ehn
(lah KAH-sah / ehl kahm-YOHN)

Use (two) men.
Usar (dos) hombres.
oo-SAHR (dohs) OHM-brehs

Get some help.
Conseguir ayuda.
kohn-seh-GEER ah-YOO-dah

Don't break the corners.
No quebrar las esquinas.
noh keh-BRAHR lahs eh-SKEE-nahs

Number of sheets

Put (twenty-five) here.
Poner (veinte y cinco) aquí.
poh-NEHR (bain-tee ee SEEN-koh) ah-KEE

Location (for laying out)

Stack it in (the house / the building).
Apilarla en (la casa / el edificio).
ah-pee-LAHR-lah ehn (lah KAH-sah / ehl eh-dee-FEE-syoh)

...on the (first / second / third) floor
...en el (primer / segundo / tercer) piso
...ehn ehl (pree-MEH-roh / seh-GOON-doh /
tehr-SEHR) PEE-soh

Drywall size (for laying out)

Use (eight foot / twelve foot) sheets here.
Usar hojas de (ocho / doce) pies aquí.
oo-SAHR OH-hahs de *(OH-choh / DOH-seh) pyehs ah-KEE*

...quarter inch / half inch / five eighths inch drywall here
...hoja de yeso de un carto de pulgada / media pulgada /
cinco octavos de pulgada aquí
...oh-hah deh YEH-soh deh oon KWAHR-toh deh
pool-GAH-dah / MEH-dyah / SEEN-koh ohk-TAH-bohs
deh pool-GAH-dah ah-KEE

Drywall type
(for laying out)

Type X for (the garage walls / the ceiling).
El tipo X es para (las paredes de garaje / el techo).
ehl TEE-poh eh-KEES ehs PAH-rah (lahs
peh-REH-dehs deh gah-RAH-hay / ehl TEH-choh)

The green drywall goes here.
La Hoja de yeso verde es para aqui.
lah OH-hah deh YEH-soh BEHR-day ehs PAH-rah ah-KEE

Loading access

Take it through (the front / the back door / the window).
A través de (la puerta de enfrente /
la purta de arás / la ventana).
ah trah-BEHS deh (lah PWEHR-tah deh ehn-FREHN-teh /
lah poo-EHR-tah deh ah-TRAHS / lah behn-TAH-nah)

Removing the
window

Take the window out.
Remover la ventana.
rrreh-moh-BEHR lah behn-TAH-nah

Don't damage (the door / the window).
No dañar (la puerta / la ventana).
noh dahn-YAHR (lah PWEHR-tah / lah behn-TAH-nah)

Laying out drywall

Put it (here / there).
Ponerla (aquí / allá).
poh-NEHR-lah (ah-KEE / ah-YAH)

Lay it down flat.
Recostarla completamente.
reh-kh-STAHR-lah kohm-pleh-tah-MEHN-teh.

Stand it against the wall.
Pararla contra la pared.
pahr-AHR-lah KOHN-trah lah pah-REHD

Supporting the load	**It is very heavy.** Es muy pesada. *ehs MOO-ee peh-SAH-dah*
	Stack it over the beam below. Apilarla sobre la viga abajo. *ah-pee-LAHR-lah SOH-breh lah BEE-gah ah-BAH-hoh*
Bracing the floor	**Brace the floor below.** Reforzar el piso debajo. *rrreh-fohr-SAHR ehl PEE-soh deh-BAH-hoh*

SCAFFOLDING

	(Don't) Use the scaffolding. (No) Usar el andamiaje. *(noh) oo-SAHR ehl ahn-dah-MYAH-heh*
Unloading / loading scaffolding	**(Unload /load) the scaffolding.** (Descargar /cargar) el andamiaje. *(dehs-kahr-GAHR / kahr-GAHR) ehl ahn-dah-MYEH-hay*
	(Unload / load) (eight) (planks / cross braces). (Descargar / cargar) (ocho) (tablones / viguetas cruzadas). *(dehs-kahr-GAHR / kahr-GAHR) (OH-choh)* *(tah-BLOH-nehs / bee-GEH-tahs kroo-SAH-dahs)*
	...frames / wheels / adjustable legs / leg extensions ...armaduras / ruedas / patas ajustables / extensiones de patas *...ahr-mah-DOO-rahs / roo-WEH-dahs /PAH-tahs* *ah-hoo-STAH-blehs / ehk-stehn-SYOH-nehs deh PAH-tahs*
Laying out scaffolding	**Lay out the scaffolding.** Extender el andamiaje. *ehk-stehn-DEHR ehl ahn-dah-MYAH-heh*
	Lay out it all the way around the house. Extenderla por toda la casa. *ehk-stehn-DEHR-lah pohr TOH-doh lah KAH-sah*

Put them every ten feet.
Poner una cada diez pies.
poh-NEHR OO-nah KAH-dah dyehs pyehs

We need more (cross braces).
Necesitamos más (viguetas cruzadas).
neh-seh-see-TAH-mohs mahs
(bee-GEH-tahs kroo-SAH-dahs)

Setting up scaffolding

Set up the scaffolding (here / there).
Armar el andamiaje (aquí / allá).
ahr-MAHR ehl ahn-dah-MYAH-heh (ah-KEE / ah-YAH)

Get somebody to help you.
Conseguir ayuda.
kohn-seh-GEER ah-YOO-dah

Make it safe.
Asegurarlo.
Ah-seh-goo-RAHR-loh

That's no good.
No está bien.
noh eh-STAH byehn.

Scaffold parts

Use the (adjustable legs / leg extensions) here.
Usar las (patas ajustables / extensiones de patas) aquí.
oo-SAHR lahs (PAH-tahs ah-hoo-STAH-blehs /
ehk-stehn-SYOH-nehs deh PAH-tahs) ah-KEE.

(Use / lock) the wheels.
(Usar / asegurar) las ruedas.
(oo-SAHR / ah-seh-goo-RAHR) lahs roo-WEH-dahs

Breaking down scaffolding

Take down the scaffolding.
Desarmar el andamiaje.
deh-sahr-MAHR ehl ahn-dah-MYAH-heh.

Load it on the truck.
Cargarlo en el camión.
kahr-GAHR-loh ehn ehl kahm-YOHN.

HANGING DRYWALL

Hanging drywall

Can you hang drywall?
¿Puede colgar hoja de yeso?
PWEH-deh kohl-GAHR OH-hah de YEH-soh

Hang the drywall.
Colgar hoja de yeso.
kohl-GAHR OH-hah de YEH-soh

Order of work

Do the ceiling first.
El techo primero.
ehl TEH-choh pree-MEH-roh

Do the walls second.
Las paredes segundo.
lahs pah-REH-dehs seh-GOON-doh

Start at the top.
Comenzar por arriba.
koh-mehn-SAHR pohr ah-RRREE-bah

**Location
(for hanging)**

Put two layers on (the ceiling / that wall).
Poner dos capas en (el techo / esa pared).
*poh-NEHR dohs KAH-pahs ehn
(ehl TEH-choh / EH-sah pah-REHD)*

**Drywall type
(for hanging)**

Type X drywall is for (the garage / the ceiling).
La hoja de yeso tipo X es para las paredes del (garaje / techo).
*lah OH-hah deh YEH-soh TEE-poh eh-KEES ehs PAH-rah
lahs pah-REH-dehs dehl (gah-RAH-heh / TEH-choh)*

The green drywall is goes here.
La hoja de yeso verde es para aquí.
lah OH-hah deh YEH-soh BEHR-day ehs PAH-rah ah-KEE.

No drywall here.
Ninuna hoja de yeso aquí.
neen-GOO-nah OH-hah deh YEH-soh ah-KEE.

Placing breaks

It must break on a (stud / joist).
El punto de unión de los extremos debe
caer sobre un (montante / viga).
ehl POON-toh deh oo-NYON deh lohs ehk-STREH-mohs
oon (mohn-TAHN-teh / BEE-gah)

Cut it around the (windows / doors).
Cortar alrededor de las (ventanas / puertas).
kohr-TAHR ahl-rrreh-deh-DOHR deh lahs
(behn-TAH-nahs / PWEHR-tahs)

Break it here. Not here.
Unirlo aquí. No aquí.
roo-NEER-loh ah-KEE. Noh ah-KEE

Measuring drywall

How (wide / tall) is it?
¿Cuán (ancho / alto) es?
kwahn (AHN-choh / AHL-toh) ehs

(Butt / hook) your tape.
(Tocar / enganchar) la cinta para medir.
(toh-KAHR / ehn-gahn-CHAHR) lah
SEEN-tah pah-rah meh-DEER

(Fouteen and one-half) inches (to it / to center).
A (catorce y media) pulgada(s) (a él / al centro).
ah (kah-TOHR-seh ee MEH-dyah)
pool-GAH-dah(s) (ah ehl / ehl SEHN-troh)

Cut it (four) inches (high / wide / long).
Cortar eso (cuatro) pulgadas (de alto /ancho / largo).
kohr-TAHR EH-soh (KWAH-troh) pool-GAH-dahs
(deh AHL-toh / AHN-choh / LAHR-goh)

Cutting drywall

Cut (this / that) piece.
Cortar (esta / esa) pieza.
kohr-TAHR (EHS-tah / EH-sah) PYEH-sah.

It is too (long / short).
Es demasiado (larga / corta).
ehs deh-mah-SYAAH-doh (LAHR-gah / KOHR-tah)

A little short is better.
Un poco corto es mejor.
oon POH-koh KOHR-toh ehs meh-HOHR

Planing the edge

Plane the edge.
Cepillar el borde.
seh-pee-YAHR ehl BOHR-deh

Cut-outs

Cut out for the (switch / outlet).
Cortar para el (interruptor / tomacorriente).
kohr-TAHR PAH-rah ehl (een-teh-rrroop-TOHR /
toh-mah-kohrrr-YEHN-teh)

…the light / the box / the medicine cabinet
…la luz / la caja / el botiquín
…lah LOOS / lah KAH-hah / ehl boh-tee-KEEN

Leave this open.
Dejar abierto.
deh-HAHR ah-BYEHR-toh

Electrical wires

Watch out for wires.
Cuidado con los cables.
kwee-DAH-doh kohn lohs KAH-blehs

Drywall behind
stair stringers

Slide it in behind the stairs.
Deslizarla detrás de la escalera.
dehs-lee-SAHR-lah deh-TRAHS deh lah eh-skah-LEH-rah

Gluing drywall

Glue it (here / there).
Pegarla (aquí / allá).
peh-GAHR-lah (ah-KEE / ah-YAH)

PATCHING HOLES

Preparation

Patch the hole.
Rellenar el hueco.
rrreh-yeh-NAHR ehl WEH-koh

Cutting back
to a stud

Cut it back to the stud.
Cortar hasta el montante.
kohr-TAHR AH-stah ehl mohn-TAHN-teh

Make a clean cut.
Hacer un corte limpio.
ah-SEHR oon KOHR-teh LEEM-pyoh

Cutting the patch

Cut a piece of drywall to fit.
Cortar un trazo de hoja de yeso para que quepa.
kohr-TAHR oon TROH-soh deh OH-hah
deh YEH-soh PAH-rah keh KEH-pah

(Nail / screw) it into place.
Clavarla / atornillarla.
klah-BAHR-lah / ah-tohrn-YAHR-lah

Mixing mud

Mix the mud.
Mezclar la masilla.
mehs-KLAHR lah mah-SEE-yah

Use the (twenty / forty-five / ninety) minute mud.
Usar masilla de (veinte / cuarenta y cinco / noventa) minutos.
oo-SAHR mah-SEE-yah deh (BAIN-teh / kwahr-EHN-tah
ee SEEN-koh / noh-BEHN-tah) mee-NOO-tohs

...more water / powder
...más agua / polvo
...mahs AH-gwah / POHL-boh

It can't have any lumps.
Sin terrones.
seen teh-RRROH-nehs

Taping

Tape the joint.
Encintar la junta.
ehn-seen-TAHR lah HOON-tah

Use the (four)-inch blade.
Usar la cuchilla de (cuatro) pulgadas.
oo-SAHR lah koo-CHEE-yah deh (KWAH-troh) pool-GAH-dahs

Let it dry completely.
Dejarla secar completamente.
deh-HAHR-lah seh-KAHR kohm-pleh-tah-MEHN-teh

Sand it.
Lijarla.
lee-HAHR-lah

Do you need (more screens / an extension handle)?
¿Necesita (más pantallas / un mango de extensión)?
*neh-seh-SEE-tah (mahs pahn-TAH-yahs /
oon MAHN-goh deh ehk-stehn-SYOHN)*

Mud it again.
Masillar otra vez.
Mah-see-YAHR OH-trah bais

■ BUTTERFLY PATCH

Cutting the hole

Make the hole (clean / rectangular).
Hacer un hueco (limpio / rectangular).
*ah-SEHR oon WEH-koh (LEEM-pyoh /
rrrehk-tahn-goo-LAHR)*

It is not necessary to cut it to a stud.
No es necesario cortar hasta el montante.
*noh ehs neh-seh-SAHR-yoh kohr-TAHR
AH-stah ehl mohn-TAHN-teh*

Cutting the patch

Cut a piece of drywall.
Cortar un pedazo de hoja de yeso.
kohr-TAHR oon peh-DAH-soh deh OH-hah deh YEH-soh

Make it four inches bigger than the hole.
Hacerlo cuatro pulgadas más grande que el hueco.
*ah-SEHR-loh KWAH-troh pool-GAH-dahs
mahs GRAHN-deh keh ehl WEH-koh*

Cut that piece, to fit the hole.
Cortar ese pedazo para que llene el hueco.
*kohr-TAHR EH-seh peh-DAH-soh
PAH-rah YEH-neh ehl WEH-koh*

On one side, leave the paper.
Dejar el papel en un lado.
deh-HAHR ehl pah-PEHL ehn oon LAH-doh

You need two inches of the paper flap all the way around.
Nececita dos pulgadas del papel restante alrededor.
neh-seh-SEE-tah dohs pool-GAH-dahs dehl
pah-PEHL rrrehs-TAHN-teh ahl-rrreh-deh-DOHR

**Mudding in
the patch**

Mud around the hole.
Masillar alrededor del hueco.
mah-see-YAHR ahl-rrreh-deh-DOHR dehl WEH-koh

Mud the patch into place.
Masillar el parche en su lugar.
mah-see-YAHR ehl PAHR-cheh ehn soo loo-GAHR

▮▮▮▮▮▮▮▮ NAILING DRYWALL

Nail all the drywall.
Clavar toda la hoja de yeso.
klah-BAHR TOH-dah lah OH-hah deh YEH-soh

Nails

Do you need more nails?
¿Necesita más clavos?
neh-see-SEE-tah mahs KLAH-bohs

Get more nails from my truck.
Busque más clavos de mi camión.
BOOS-keh mahs KLAH-bohs deh mee kahm-YOHN

Nailing schedule

Nail it every (eight / twelve) inches.
Clavarlo cada (ocho / doce) pulgadas.
klah-BAHR-loh KAH-dah (OH-choh / DOH-seh)
pool-GAH-dahs

Setting the nail

Sink the head below flush.
Hundir la cabeza por debajo del punto a ras.
hoon-DEER lah kah-BEH-sah pohr deh-BAH-ho
dehl POON-toh ah rahs

Missed nails

When you miss the stud, pull the nail.
Cuando no le de, saque el clavo.
KWAHN-doh noh leh deh, SAH-keh ehl KLAH-boh

Is it complete?
¿Está completo?
EH-stah kohm-PLEH-toh

SCREWING DRYWALL

Screw the drywall.
Atornillar la hoja de yeso.
ah-tohrn-YAHR lah OH-hah deh YEH-soh

Screws

Do you need more screws?
¿Necesita más tornillos?
neh-seh-SEE-tah mahs tohr-NEE-yohs

The screws are in (my / the truck).
Los tornillos están en (mi / el camión).
*lohs tohr-NEE-yohs EH-stahn ehn
(mee / ehl kahm-YOHN)*

Screw gun

Do you have a screw gun?
¿Tiene un destornillador eléctrico?
t-YEHN-eh oon deh-stohrn-YAH-dohr eh-LEHK-tree koh

The screw gun is in (my / the truck).
El destornillador eléctrico está en (mi / el camión).
*ehl Eye oon deh-stohrn-YAH-dohr eh-LEHK-tree
koh EH-stah ehn (mee / ehl kahm-YOHN)*

Screwing

Screw it every (eight / twelve) inches.
Atornillar cada (ocho / doce) pulgadas.
*ah-tohrn-YAHR KAH-dah (OH-choh /
DOH-seh) pool-GAH-dahs*

The head must be below flush.
La cabeza debe quedar por debajo del punto a ras.
*lah kah-BEH-sah DEH-beh keh-DAHR pohr
deh-BAH-ho dehl POON-toh ah rahs*

Is it complete?
¿Está completo?
eh-STAH kohm-PLEH-toh

████████████ CORNER BEAD

Install the corner beads.
Instalar los angulares de esquina.
een-stah-LAHR lohs ahn-goo-LAHR-ehs deh eh-SKEE-nah

Type of corner bead

Use (bull nose / standard).
Usar el (redondo / estándar).
oo-SAHR (reh-DOHN-doh / ehs-TAN-dahr)

Use (metal / flexible) corner beads.
Usar los angulares de esquina (metálicos / flexibles).
oo-SAHR lohs ahn-goo-LAHR-ehs deh
eh-SKEE-nah (meh-TAH-lee-kohs / flek-SEE-blehs)

Corner bead location

Put corner bead there.
Poner angular de esquina allá.
poh-NEHR ahn-goo-LAHR deh eh-SKEE-nah ah-YAH

No corner bead here.
Ningún angular de esquina aquí.
neen-GOON ahn-goo-LAHR deh eh-SKEE-nah ah-KEE

Cutting corner bead

It's too (long / short).
Es demasiado (largo / corto).
ehs deh-mah-SYAH-doh (LAHR-goh / KOHR-toh)

Nailing corner bead

Nail it every (twelve) inches.
Clavarlo cada (doce) pulgadas.
klah-bahr-loh KAH-dah (DOH-seh) pool-GAH-dahs

████████████ TAPING DRYWALL

Taping drywall

Do you know how to tape drywall?
¿Sabe encintar la hoja de yeso?
SAH-beh ehn-seen-TAHR lah OH-hah deh YEH-soh

Tape the drywall.
Encintar la hoja de yeso.
ehn-seen-TAHR lah OH-hah deh YEH-soh

Firetaping	**Only firetape the (garage / basement).** Poner cinta anti-incendia sólo en el (garaje / sótano). *poh-NEHR SEEN-tah ahn-tee een-SEHN-deeo SOH-loh* *ehn ehl (gah-RAH-heh / soh-TAHN-oh)*
Mudding the walls	**Mud (the joints / the nails).** Masillar (las juntas / los clavos). *mah-see-YAHR (lahs HOON-tahs / lohs KLAH-bohs)* **Use a (four / six / ten / twelve)-inch knife.** Usar un cuchillo de (cuatro / seis / diez / doce) pulgadas. *oo-SAHR oon koo-CHEE-yoh deh (KWAH-troh /* *sais / dyehs / DOH-seh) pool-GAH-dahs* **Use (more / less) mud.** Usar (más / menos) masilla. *oo-SAHR (mahs / MEH-nohs) mah-SEE-yah* **That mud is trash.** Esa masilla es basura. *EH-sah mah-SEE-yah ehs bah-SOO-rah*
Applying tape	**Apply the tape.** Aplicar la cinta. *ah-plee-KAHR lah SEEN-tah* **Put tape here.** Poner cinta aquí. *poh-NEHR SEEN-tah ah-KEE* **It must be smooth.** Debe quedar lisa. *DEH-beh keh-DAHR LEE-sah* **Feather it in.** Allenar la pared. *ah-yeh-NAHR lah pah-REHD*
Drying mud	**Let it dry.** Dejarla secar. *deh-HAHR-lah seh-KAHR*

Is it dry?
¿Está seca?
EH-stah SEH-kah

It is (not) dry.
(No) está seca.
(noh) eh-STAH SEH-kah

Sanding

Sand the (walls / ceilings) (again).
Lijar (las paredes / los techos) (otra vez).
lee-HAHR (lahs pah-REH-dehs /
lohs TEH-chohs) (OH-trah bais)

Do you need an (extension handle / sanding pads)?
¿Necesita (un mango de extensión / hojas de lija)?
neh-seh-SEE-tah oon (MAHN-goh deh
ehk-stehn-SYOHN / OH-hahs deh LEE-hah)

It must look good.
Debe lucir bien.
DEH-beh loo-SEER byehn

Dust mask /
respirator

Wear a (dust mask / respirator).
Usar (máscara anti-polvo / respirador).
oo-SAHR (mahs-KAHR-ah AHN-tee
POHL-boh / rrreh-speer-ah-DOHR)

▰▰▰▰▰▰ TEXTURING

Shoot the (walls / ceilings).
Rociar (las paredes / los techos).
rrroh-SYAHR (lahs pah-REH-dehs / lohs TEH-chohs)

Describing the
texture

The (walls / the ceilings) should have a smooth surface.
(Las paredes / los techos) deben tener una superficie.
(las pah-REH-dehs / lohs TEH-chohs) deh-BEHN
teh-NEHR oo-NAH soo-pehr-FEE-syeh.

...orange peel / skip trowel
...cáscara de naranja / paleta
...kahs-KAH-rah deh nah-RAHN-hah / pah-LEH-tah

…knock-down / light / medium / heavy
…derribar / ligero / mediano / grueso
…deh-rrree-BAHR / lee-GEH-roh /
meh-DYAH-noh / groo-WEH-soh

Mask (the windows / the wood / the fixtures).
Enmascarar (las ventanas / la madera / las instalaciones).
ehn-mahs-kah-RAHR (lahs behn-TAH-nahs /
lah mah-DEH-rah / las een-stah-lah-SYON-ehs)

Do you need more (plastic / masking tape)?
¿Necesita más (plástico / cinta de enmascarar)?
neh-seh-SEE-tah mahs (PLAH-stee-koh /
SEEN-tah deh ehn-mahs-kah-RAHR)

Cover the floor.
Cubrir el piso.
koo-BREER ehl PEE-soh

Spread a drop cloth.
Extender una tela protectora.
ehk-stehn-DEHR OO-nah TEH-lah proh-tehk-TOH-rah

It's in (my truck / the house).
Está en (mi camión / la casa).
EH-stah ehn (mee kahm-YOHN / lah KAH-sah)

Set up the compressor and hose.
Preparar el compresor y la manguera.
preh-pahr-RAHR ehl kohm-preh-SOHR ee lah mahn-GEH-rah

Get the hopper and mud.
Conseguir la tolva y la masilla.
kohn-seh-GEER la TOHL-bah ee lah mah-SEE-yah

Add more (water / mud).
Aregar más (agua / masilla).
ah-greh-GAHR mahs (AH-gwah / mah-SEE-yah)

A (little / lot) more.
(Un poco / mucho) más.
Oon (POH-koh / MOO-choh) mahs

TELEPHONE

Get my phone.
Traer mi teléfono.
trah-EHR mee teh-LEH-foh-noh

It's in (my truck / the trailer).
Está en (mi camión / el trailer).
EH-stah ehn (mee kahm-YOHN / ehl trailer)

FIRST AID KIT

Get the first aid kit.
Traer el botiquín de primeros auxilios.
trah-ER ehl boh-tee-KEEN de pree-MEHR-ohs ahk-SEEL-yohs.

It's in my (truck / the trailer).
Está en (mi camión / el tráiler).
EH-stah ehn (mee kahm-YOHN / ehl trailer)

FALL INJURY

Don't move him.
No moverlo.
noh moh-BEHR-loh

Cover him.
Cubrirlo.
koo-BREER-loh

HEAD INJURY

What is your name?
¿Cómo se llama?
KOH-moh seh YAH-mah

Do you know where you are?
¿Sabe dónde está?
SAH-beh DOHN-deh eh-STAH

Keep him quiet.
Mantenerlo calmado.
mahn-teh-NEHR-loh kahl-MAH-doh

Talk to him.
Hablarle.
ah-BLAHR-leh

CUTS

Apply pressure here.
Aplicar presión aquí.
ahp-lee-KAHR preh-SYOHN ah-KEE

Put this around his (leg / arm).
Poner esto alrededor de su (pierna / brazo).
poh-NEHR EH-stoh ahl-rrreh-deh-DOHR
deh soo (PYEHR-nah / BRAH-soh)

Hold it here.
Sostenerlo aquí.
soh-steh-NEHR-loh ah-KEE

Keep it tight.
Mantenerlo apretado.
mahn-teh-NEHR-loh ah-preh-TAH-doh

NAIL GUN INJURIES

(Don't) pull the nail.
(No) sacar el clavo.
(noh) sah-KAHR ehl KLAH-boh

Cut the plywood here.
Cortar la madera terciada aquí.
kohr-TAHR lah mah-DEH-rah tehr-see-AH-dah ah-KEE

EYE INJURY (NAIL IMPALEMENT)

Don't pull the nail.
No sacar el clavo.
no sah-KAHR ehl KLAH-boh

Immobilize it with tape.
Inmovilizarlo con cinta.
ee-moh-bee-lee-SAHR-loh kohn SEEN-tah

Hold him still.
Mantenerlo quieto.
mahn-teh-NEHR-loh KYEH-toh

EYE INJURY (IRRITANT)

Don't rub it.
No frotarlo.
noh froh-TAHR-loh

Come with me.
Venga conmigo.
BEHN-gah kohn-MEE-gah

Use the eyewash.
Usar el colirio.
oo-SAHR ehl koh-LEER-yoh

Watch, do it like this.
Mire, así.
MEE-reh, ah-SEE

ELECTRICAL INJURY

Don't touch him.
No tocarlo.
noh toh-KAHR-loh

Unplug the cord.
Desconectar el cable.
dehs-koh-nehk-TAHR ehl KAH-bleh

SHOCK

Elevate his legs a little.
Elevar sus piernas un poco.
eh-leh-BAHR soos PYEHR-nahs oon POH-koh

Cover him.
Cubrirlo.
koo-BREER-loh

APPENDIX A

Measurement conversion chart

Conversion	Procedure	Example
°F to °C	Subtract 32 and multiply by 0.555	60 °F = (60 - 32) x 0.56 15.68 °C = 28 x 0.555
feet to m	Multiply by 0.305	2 feet = 2 x .305 m = 0.61 m
psf (pounds/foot^2) to kg/m^2	Multiply by 4.88	25 psf = 25 x 4.88 kg/m^2 = 122 kg/m^2
psi (pounds/inch^2) to kg/m^2	Multiply by 0.07	1310 psi = 1310 x 0.07 kg/m^2 = 91.7 kg/m^2
inches to cm	Multiply by 2.54	3 inches = 3 x 2.54 cm = 7.63 cm
inches to mm	Multiply by 25.4	3 inches = 3 x 25.4 mm = 76.3 mm
yard^3 to m^3	Multiply by 0.765	5 yard^3 = 5 x 0.765 m^3 = 3.825 m^3

APÉNDICE A

Gráfico de conversión de medidas

Conversión	Procedimiento	Ejemplo
°F a °C	Restar 32 y multiplicar por 0.555	60 °F = (60 - 32) x 0.56 15.68 °C = 28 x 0.555
pies a m	Multiplicar por 0.305	2 pies = 2 x .305 m = 0.61 m
psf (libras/pie^2) a kg/m^2	Multiplicar por 4.88	25 psf = 25 x 4.88 kg/m^2 = 122 kg/m^2
psi (libras/pulg^2) a kg/m^2	Multiplicar por 0.07	1310 psi = 1310 x 0.07 kg/m^2 = 91.7 kg/m^2
pulgadas a cm	Multiplicar por 2.54	3 pulg. = 3 x 2.54 cm = 7.63 cm
pulgadas a mm	Multiplicar por 25.4	3 pulg. = 3 x 25.4 mm = 76.3 mm
yarda^3 a m^3	Multiplicar por 0.765	5 yardas^3 = 5 x 0.765 m^3 = 3.825 m^3

INDEX

Page numbers in *italics* refer to illustrations

Foam insulation, 149, 153
Footing dimensions, 21–22
Form liners, 28
Form ties, 29–30
Forms (concrete)
  heavy, 30–31
  setting of
    for footings and slab, 19, 21–23
    for stairways, 32
    for walls, 25–30
  soaking of, 37
Forms (paper)
  filling out of, 2
  immigration, 3
  tax, 3
Framing
  roof, *see* Roof framing
  safety terms, 8–9
  walls, *see* Framing walls
Framing members, *48*
Framing walls
  bracing walls, 55
  clearing the floor, 47
  crowning studs, 49–50
  labeling of wall plates, 47, 49
  laying out walls, 49
  layout, 49
  nailing, 50
  nailing off corners, 55
  ridge beam pockets, 51–52
  separating wall plates, 50
  standing walls, 54
  temporary braces, 54–55
  top plate, 50–51
Freeze protection, 44

G

Gable studs, 106
Gable vent, 106

Galvanized nails, 123
Gang cutting, 46
Glasses, safety, 10
Gluing
  drywall, 161
  joints, 112
  sheathing, 72
Grading
  driveways, 32–33
  ramps, 31
  stairway, 31
Gravel, spreading of, 132
Guardrails, 8
Guiding heavy forms, 30
Gutter, 143

H

Hammer marks, 112
Hangers
  flush, 69
  laying out joists into, 63
  nailing of, 69
  pulling layout of, 69
Hanging
  backing, 68, 90
  drywall, 159–161
  jacks, 104
Hard hat, 5
Hardware
  beam, 61
  embedded, 40
Harness, 6
Head
  cutting, 45–47
  shimming the, 80
Head injury, 171–172
Heating ducts, insulating of, 149–150
Height
  scaffolding, 83, 97

Nail gun, 15, 79, 123, 139
Nail gun injuries, 172
Nailing
  backing, 68, 105
  braces, 57, 108
  catwalks, 89
  checking of, 119
  corner bead, 166
  doors, 79
  drywall, 164–165
  fascia, 112
  felt, 132
  flanges, 77
  flush-nailing, 59
  joists, 67, 122
  rim, 64
  schedule for, 50–51, 60, 65, 88,
      101, 104, 119, 136, 141, 164
  shear panel, 59
  sheathing, 74
  soffits, 120, 123
  studs, 50
  top plate, 51
  valley flashing, 129
Nailing off
  corners, 55, 73
  roof, 118
  top plate, 56
Name, 2
Not hiring, 2
Nouns, xi
Number, 2
Nutting down, 56

O
Offsetting breaks, 136
Oil
  changing of, 17
  checking of, 14

  nail gun, 15
Open valley, 137
Operating the window, 76
Orange vest, 37
Overhang
  interlayment, 141
  joists, 63, 66–67
  marking lengths of, 108
  shingles, 133
Overlap
  cutting of, 117–118
  interlayment, 139
  planking, 108
  rebar, *34,* 35
  top plate, 51
  underlayment, 131

P
Paper
  exposure of, 131
  installing of, 130–131
Paperwork, 2–3
Patching holes, 161–163
Paycheck, 4
Payday, 1–2
Payroll, 4
Perimeter, 74
Pin locations, 57
Pipes, insulating of, 149
Placing
  sheathing, 72–73
  windows, 75–76, 78
Planking, 108
Planks, 83, 97
Plastic, 43, 150
Plugs, 9
Plumb/plumbing
  doors, 78–79
  forms, 29